Dear Joe,

Thanks for continually fostering my research since I arrived at CIIS. Without your generous sabbaticals and other support, I couldn't have done it.

Best,
Doug

Psychology of Space Exploration

Contemporary Research in Historical Perspective

Edited by Douglas A. Vakoch

The NASA History Series
National Aeronautics and Space Administration
Office of Communications
History Program Office
Washington, DC
2011

NASA SP-2011-4411

Library of Congress Cataloging-in-Publication Data

Psychology of space exploration : contemporary research in historical perspective / edited by Douglas A. Vakoch.
p. cm. -- (NASA history series)
1. Space psychology. 2. Space flight--Psychological aspects. 3. Outer space--Exploration. 4. Space sciences--United States. I. Vakoch, Douglas A. II. United States. National Aeronautics and Space Administration.
RC1160.P79 2009
155.9'66--dc22

2009026665

For sale by the Superintendent of Documents, U.S. Government Printing Office
Internet: bookstore.gpo.gov   Phone: toll free (866) 512-1800;   DC area (202) 512-1800
Fax: (202) 512-2104 Mail: Stop IDCC, Washington, DC 20402-0001

To Julie and Len

Table of Contents

Foreword vii
Acknowledgments ix

Chapter 1. Introduction: Psychology and the U.S. Space Program 1
    Albert A. Harrison, Department of Psychology, University of California, Davis
    Edna R. Fiedler, National Space Biomedical Research Institute, Baylor College
        of Medicine

Section I: **Surviving and Thriving in Extreme Environments**

Chapter 2. Behavioral Health 17
    Albert A. Harrison, Department of Psychology, University of California, Davis
    Edna R. Fiedler, National Space Biomedical Research Institute, Baylor College
        of Medicine

Chapter 3. From Earth Analogs to Space: Getting There from Here 47
    Sheryl L. Bishop, Department of Preventive Medicine and Community Health
        and School of Nursing, University of Texas Medical Branch

Chapter 4. Patterns in Crew-Initiated Photography of Earth from the
ISS—Is Earth Observation a Salutogenic Experience? 79
    Julie A. Robinson, Office of the ISS Program Scientist, National Aeronautics
        and Space Administration (NASA) Johnson Space Center (JSC)
    Kelley J. Slack, Behavioral Health and Performance Research, Wyle Laboratories
    Valerie A. Olson, Department of Anthropology, Rice University
    Michael H. Trenchard, Image Science and Analysis Laboratory, Engineering and
        Science Contract Group (ESCG), NASA JSC
    Kimberly J. Willis, Image Science and Analysis Laboratory, ESCG, NASA JSC
    Pamela J. Baskin, Behavioral Health and Performance Research, Wyle Laboratories
    Jennifer E. Boyd, Department of Psychiatry, University of California, San
        Francisco; and San Francisco Veterans Affairs Medical Center

Psychology of Space Exploration

Section II: Managing Interpersonal Conflict in Space

Chapter 5. Managing Negative Interactions in Space Crews: The Role of Simulator Research 103
  Harvey Wichman, Aerospace Psychology Laboratory, Claremont McKenna
    College and Claremont Graduate University

Chapter 6. Gender Composition and Crew Cohesion During Long-Duration Space Missions 125
  Jason P. Kring, Department of Human Factors and Systems, Embry-Riddle
    Aeronautical University
  Megan A. Kaminski, Program in Human Factors and Applied Cognition, George
    Mason University

Section III: Multicultural Dimensions of Space Exploration

Chapter 7. Flying with Strangers: Postmission Reflections of Multinational Space Crews 143
  Peter Suedfeld, Department of Psychology, University of British Columbia
  Kasia E. Wilk, Youth Forensic Psychiatric Services Research and Evaluation
    Department, Ministry of Children and Family Development
  Lindi Cassel, Department of Occupational Therapy, Providence Health Care

Chapter 8. Spaceflight and Cross-Cultural Psychology 177
  Juris G. Draguns, Department of Psychology, Pennsylvania State University
  Albert A. Harrison, Department of Psychology, University of California, Davis

Afterword. From the Past to the Future 195
  Gro Mjeldheim Sandal, Department of Psychosocial Science, University of Bergen
  Gloria R. Leon, Department of Psychology, University of Minnesota

About the Authors 205
Acronyms and Abbreviations 219
The NASA History Series 221
Subject Index 235
Authors Cited 249

Foreword

Each month, the cover of *Monitor on Psychology*, a magazine sent to over one hundred thousand members of the American Psychological Association, reflects intriguing new areas of interest to psychologists who work as researchers, clinicians, consultants, and teachers. The importance of human adaptation to space for contemporary psychologists is suggested by the cover of the March 2008 *Monitor*, which featured an astronaut drifting in space, with the tranquil blue Earth in the background and the caption "Deep Space Psych" below.

At one level, the essays in this volume provide an overview and synthesis of some of the key issues in the psychology of space exploration, as well as a sampling of highly innovative empirical research. The characteristic that most clearly sets this collection apart from others, however, is the depth with which the authors have engaged the *history* of the psychology of space exploration.

All psychologists are familiar with the importance of engaging past research and theory while conducting literature reviews in preparation for designing and interpreting new studies. But the contributors to this collection have done much more. They have crafted essays that will be of obvious value to psychologists, psychiatrists, and other behavioral researchers. At the same time, these authors have created a collection with the promise to promote a greater dialogue between psychological researchers and both historians of space exploration and historians of psychology.

Psychologists and historians have quite different criteria for good scholarship and for communicating their findings. These differences make the essays in this volume—meaningful and accessible even to those not formally trained in psychologists' methodologies and mindsets—all the more impressive. With the increasing specialization and isolation of academic disciplines from one another over the past century, these essays serve as a prototype for a broader attempt to bridge the gap between the two cultures of science and the humanities that C. P. Snow identified almost a half century ago—quite fittingly for us, near the beginning of the Space Age. Let us hope that as we prepare once again to send astronauts beyond Earth's orbit, we can do so with the guidance of others equally open to seeing beyond their own specialties.

Acknowledgments

Without the intellectual leadership of Albert Harrison, this book would never have come into existence, and it could not have been completed in such a timely manner. His contributions will be evident in the three chapters he has coauthored; invisible is his extensive work recruiting other contributors, reviewing chapters, and providing last-minute assistance more times than I care to remember. Much more important to me, however, is Al's ongoing friendship.

Over the past decade, many colleagues from the SETI Institute have shared with me their insights about the societal and educational impact of space exploration—especially John Billingham, Edna DeVore, Frank Drake, Andrew Fraknoi, John Gertz, Chris Neller, Tom Pierson, Karen Randall, Seth Shostak, and Jill Tarter. More recently, I warmly acknowledge the administration, faculty, staff, and students of the California Institute of Integral Studies (CIIS), especially for support from Katie McGovern, Joseph Subbiondo, and Judie Wexler. The work of editing this volume was made possible through a generous sabbatical leave from my other academic responsibilities at CIIS. In addition, I thank Harry and Joyce Letaw, as well as Jamie Baswell, for their intellectual and financial contributions to promoting the societal aspects of space exploration.

Among the organizations that have fostered discussions on the topics in this volume, I especially want to recognize the International Academy of Astronautics (IAA) and the American Psychological Association (APA). Several of the chapters in this volume are elaborations of papers first presented at the APA's 115th Annual Convention, held in San Francisco in August 2007.

For his openness to considering a new topic for the NASA History Series, I thank Steve Dick; I am also grateful to him and to Steve Garber for leading such a thorough and helpful review process and for moving this volume into production so efficiently.

In the Communications Support Services Center at NASA Headquarters, Lisa Jirousek copyedited the manuscript, Christopher Yates designed the layout, Stacie Dapoz and George Gonzalez proofread the layout, and Hanta Ralay and Tun Hla handled the printing. Supervisors Gail Carter-Kane, Cindy Miller, Michael Crnkovic, and Tom Powers oversaw the overall process. Thanks are due to all of these fine professionals.

Finally, I thank the contributors and reviewers of the essays that appear in this volume. By taking seriously the importance of history for contemporary psychological research, they have opened new possibilities for interdisciplinary collaborations in the future.

Douglas A. Vakoch
Mountain View and San Francisco, California

Chapter 1
Introduction: Psychology and the U.S. Space Program

Albert A. Harrison
>Department of Psychology
>University of California, Davis

Edna R. Fiedler
>National Space Biomedical Research Institute
>Baylor College of Medicine

ABSTRACT

Astronauts live and work in highly unusual and challenging environments where they must withstand multiple stressors. Their abilities to maintain positive psychological outlooks and good interpersonal relations are crucial for personal well-being and mission success. From the inception of the space program, psychologists, psychiatrists, human factors experts, and other professionals have warned that the psychological stressors of space should be treated as a risk factor and that the risk would increase as missions involved larger, more diversified crews undertaking increasingly long flights. Thus, they called for research leading to the development and application of effective countermeasures. Although psychology played a significant role at the inception of the space program, for many years thereafter certain areas of psychology all but disappeared from NASA. Interest in psychosocial adaptation was rekindled in the mid-1990s when astronauts joined cosmonauts on the Russian space station *Mir*. NASA's recognition of the field of behavioral health and its links to performance opened the door to many kinds of research that were formerly overlooked. Focusing on the underutilized areas of personality and social psychology, the chapters that follow discuss psychology's struggle for acceptance, the history of astronaut selection and psychological support, the use of analog environments and simulators for research and training, space tourism, the psychological rewards of viewing Earth from space, crew composition and group dynamics, and cross-cultural aspects of international missions. This book concludes with a summary, integration, and evaluation of the role of psychology in space exploration.

INTRODUCTION

"Once, I was evaluating astronaut applicants" says psychiatrist Nick Kanas. "I asked them to give me some examples of things that might cause stress." One applicant, a test pilot, recalled the time he was flying an experimental aircraft and it spun out of control. As the plane spiraled down, he took out his manual, calmly thumbed through it, and figured out how to pull the plane to safety. "His ability to temporarily control his emotions was striking," laughs Kanas.[1]

Fully aware of astronauts' remarkable strengths, Kanas also knows that many physical and psychological stressors can pose risks to safety, performance, and quality of life.[2] Some of these stressors are associated with flight: riding atop a rocket; rapid acceleration and deceleration; somewhat primitive living conditions; isolation from family and friends; close confinement with other people; and the ever-present specter of a collision, system failure, or other disaster. Other types of stressors come from the astronaut's career. From the earliest days of the space program, astronauts have served as societal exemplars, living under intense public scrutiny; carried heavy workloads on Earth as in space; and undergone prolonged absences from home for training, flight, and other purposes. They must withstand the typical hassles of trying to succeed within large bureaucracies, worry over flight assignments, and readjust to their families when they return to Earth.[3] J. Kass, R. Kass, and I. Samaltedinov describe how some of this may seem to an astronaut:

---

1. "How Astronauts Get Along," Science@NASA, 21 October 2002, available at http://science.msfc.nasa.gov/headlines/y2002/21oct_getalong.htm (accessed 29 March 2008).

2. N. Kanas, "Psychosocial Factors Affecting Simulated and Actual Space Missions," *Aviation, Space, and Environmental Medicine* 56, no. 8 (August 1985): 806–811; N. Kanas, "Psychosocial Support for Cosmonauts," *Aviation, Space, and Environmental Medicine* 62, no. 4 (August 1991): 353–355; N. Kanas, V. P. Salnitskiy, J. B. Ritsher, V. I. Gushin, D. S. Weiss, S. A. Saylor, O. P. Kozerenko, and C. R. Marmar, "Human Interactions in Space: ISS vs. Shuttle/Mir," *Acta Astronautica* 59 (2006): 413–419.

3. W. E. Sipes and S. T. Vander Ark, "Operational Behavioral Health and Performance Resources for International Space Station Crews and Families," *Aviation, Space, and Environmental Medicine* 76, no. 6, sec. II (June 2005): B36–B41.

Introduction: Psychology and the U.S. Space Program

He knows he has been trained and put into space at great cost and effort, and he has a limited amount of time, especially during a short shuttle mission, to perform the tasks set out for him, efficiently. The precious data of the scientists on the ground, who have dedicated many years for this experiment, can be lost, the equipment can be damaged in such a way that it cannot be repaired in space, or worse still, his blunder can affect the safety of life on the spaceship. Even if such drastic errors are seldom, he is nevertheless under great stress—he has to get the work done quickly, so that the next scheduled event can take place as planned. This kind of stress affects him not only as an individual, but as a member of a team: His peers are watching him, and he knows full well, not only will any mistakes made affect their work as well, but he fails in their eyes in a similar manner as a member of a sports team, whose error can affect the success of the team as a whole.[4]

This book discusses selected topics in the psychology of space exploration. In this and the following chapters, we and other contributors address the changing role of psychology within the U.S. space program, review the history of astronaut selection and training, and describe the evolution of techniques for providing astronauts and cosmonauts with psychological support. Contributing authors explain why and how psychologists use space-reminiscent settings (such as undersea habitats and polar outposts) for research and training purposes. They trace the not-always-smooth course of the diversification of the astronaut corps from a homogenous collection of white, male test pilots to a more heterogeneous group including women and minorities. They tell about group dynamics and teamwork, as well as occasional friction between crews in flight and people in Mission Control. One of the most dramatic changes over 50 years of crewed flight has been the transition from fiercely competitive national space programs to collaborative efforts with international crews, so cultural issues are discussed. Over the past 50 years, space missions have changed, and so have salient behavioral issues and opportunities for psychologists.

---

4. J. Kass, R. Kass, and I. Samaltedinov, "Psychological Problems of Man in Space: Problems and Solutions," *Acta Astronautica* 36 (1995): 657–660.

How has psychology fared in the U.S. space program? In his presidential address to the Division of Engineering Psychology on 4 September 1961, Walter F. Grether affirmed psychology's crucial role in the newly initiated conquest of space, noting that psychologists of that day were responding with creativity and vigor to the enormous behavioral challenges.[5] Looking back over the history of aviation, Grether remarked that despite a few contributions to military aviation in World War I, for roughly 35 years after the Wright brothers' initial flight at Kitty Hawk, aviation and psychology pretty much went separate ways. Then, beginning with research to benefit civilian aviation in the late 1930s and followed by a powerful military program in World War II, aviation psychology became prominent and influential. "How much different the role of psychology has been in man's early ventures into space!" Grether wrote.[6] Psychological testing, he continued, was prominent in the selection of the initial seven Mercury astronauts, and beyond selection psychologists were productively engaged in vehicle design, training, task design, and workload management.

Grether pointed to four areas for future research: moving about the interior of spacecraft (once they became large enough for this to occur), conducting extravehicular activities (EVAs) or "spacewalks," performing rendezvous, and living and working under conditions of prolonged isolation and confinement. Highly optimistic about America's future in space, Grether foresaw a strong continuing partnership between psychology and space exploration. One of his few notes of pessimism—that it would not be possible to use the science fiction writer's rocket gun to move from place to place during EVAs—would soon be proven wrong. Beyond providing psychologists with new opportunities for employment and research support, he felt, space exploration would open new frontiers of knowledge and stimulate thinking about new problems that would lead to new theories, hypotheses, and methods.

Nearly three decades later, participants at the 30th anniversary of the 1959 founding of the International Ergonomics Association might conclude that Grether was right. In the field generally known as human factors in the U.S. and ergonomics in the United Kingdom (U.K.) and Europe, human factors specialists are interested in the scientific problems of experimental psychology, anatomy, and physiology

---

5. W. F. Grether, "Psychology and the Space Frontier," *American Psychologist* 17, no. 2 (February 1962): 92–101.
6. Ibid., pp. 92–93.

applied to human work. Classically, human factors addresses people's interaction with physical environments in work settings, but the interests of human factors specialists have broadened over the years.[7]

In this 1989 presidential address to the association, Alphonse Chapanis could point with pride to rapidly accumulating accomplishments everywhere in the field.[8] Floods of data were appearing in area after area of human activity (work, transportation, leisure-time pursuits), and it was no longer possible to keep abreast of the latest journals and books. The hottest topic of 1989 was computers: how they had revolutionized society, how they spread beyond science and business and were embraced by everyday people, and how they could be humanized through the design of displays and controls. Certainly, much was left to be done—over the lifetime of the association, 71 major railroad disasters had claimed 5,059 persons; 192 major aircraft accidents had killed over 20,000 people; and, in the previous 10 years alone, there had been thousands of nuclear "mishaps," including prominent events at Chernobyl and Three Mile Island. Still, Chapanis's theme was that ergonomics had "come a long way, baby," and that the biggest stimulus for this was America's forays into space:

> Space flights have become so commonplace and so much is known about human performance in space that it is hard to remember the thousands of analyses, studies, and experiments that were done to pave the way for man's leap into these hostile and unknown regions. There were problems of vehicle design involving exotic displays and controls. There were problems of vibration, of g-forces, and of weightlessness that had to be explored and solved. For extravehicular activity an entire self-contained environment had to be designed for astronauts and cosmonauts. Torqueless tools had to be designed for use by men who were floating freely and encumbered by space suits with limited mobility. There were problems of nutrition, waste disposal, and work-rest cycles. Nor

---

7. D. Meister, *Conceptual Aspects of Human Factors* (Baltimore, MD: Johns Hopkins University Press, 1989).

8. A. Chapanis, "The International Ergonomics Association: Its First 30 Years," *Ergonomics* 33, no. 3 (1990): 275–282.

can we forget the problems of selection, training, and simulator design.... Our leap into space was a significant accomplishment of the past 30 years and the ergonomic findings that helped bring it about have enriched our profession in countless ways.[9]

But other assessments of psychologists' contributions to the U.S. space program were less triumphant. In 1975, Robert L. Helmreich expressed pessimistic views of applying psychology in new areas, stating that prospective customers often respond with profound indifference.[10] In 1983, he elaborated on how data relating to personality and social psychology were underused by the U.S. space program, which (as we shall see in chapter 2) he considered in contrast to robust use in the Soviet program.[11] In a 1987 conference cosponsored by NASA and the National Science Foundation, psychologist and management consultant Philip R. Harris observed that

[a]lthough NASA has been forthright about medical and biological insights gained from previous spaceflights... the agency has been hesitant on studying or releasing information on the psychosocial experience of its personnel in space. Generally, NASA has limited the access to astronauts by social science researchers, even by its own psychiatrists and psychologists; the agency has failed to capitalize on the data it collected that could improve spaceflight and living for others to follow.[12]

In the early 1990s, outgoing flight surgeon and psychiatrist Patricia Santy concluded that despite an initial flurry of interest, behavioral research all but disap-

---

9. Ibid., pp. 276–277.

10. R. L. Helmreich, "Applied Social Psychology: The Unfulfilled Promise," *Personality and Social Psychology Bulletin* 1 (1975): 548–561.

11. R. L. Helmreich, "Applying Psychology to Outer Space: Unfulfilled Promises Revisited," *American Psychologist* 38 (1983): 445–450.

12. P. R. Harris, "Personnel Deployment Systems: Managing People in Polar and Outer Space Environments," in *From Antarctica to Outer Space: Life in Isolation and Confinement*, ed. A. A. Harrison, Y. A. Clearwater, and C. P. McKay (New York: Springer, 1990), pp. 77–78.

peared from NASA.[13] For years, she wrote, psychology played a minimal role in astronaut selection, and because the assessment of individual astronaut performance was prohibited, it was not possible to collect normative data for test validation and other purposes. She characterized the application of psychology to space as running 20 to 30 years behind most areas of medicine and identified formidable organizational barriers to psychology within NASA. Joseph V. Brady, whose research on primate behavior in spaceflight dates back to the 1950s, states that following John Glenn's flight, there was a dearth of in-flight behavioral experiments.[14] Brady characterizes this as a 30-year hiatus in psychological health research for NASA, a gap that he thought must come to an end given NASA's vision for humans in space. Peter Suedfeld cuts to the heart of the matter: "Through most of NASA's existence, the behavioral sciences have been barely visible on the agency's horizon."[15]

How can we reconcile such pessimistic views with the optimistic assessments of Grether and Chapanis? Robert Helmreich's point was that, generally, those disciplines that are rooted in biology, engineering, and experimental psychology have found greater acceptance within the space program than disciplines rooted in personality, social, and organizational psychology. Lawrence Palinkas, an anthropologist who has developed an enviable record of hands-on research experience in unusual environments, organized these issues in long-term spaceflight into three "domains": the individual domain (stress and coping), the group dynamics domain (social interaction and intergroup relations), and the organizational domain (management, organizational culture, and behavior).[16]

From the beginning, physicians, psychologists, and their allies advocated strong behavioral research programs in NASA. Margaret A. Weitekamp points out how interest in high-altitude flight in the 1930s initiated research that evolved into aero-

---

13. P. A. Santy, *Choosing the Right Stuff: The Psychological Selection of Astronauts and Cosmonauts* (Westport, CT: Praeger, 1994).

14. J. V. Brady, "Behavioral Health: The Propaedeutic Requirement," *Aviation, Space, and Environmental Medicine* 76, no. 6, sect. II (June 2005): B13–B23.

15. P. Suedfeld, "Invulnerability, Coping, Salutogenesis, Integration: Four Phases of Spaceflight Psychology," *Aviation, Space, and Environmental Medicine* 76, no. 6, sect. II (June 2005): B61.

16. L. Palinkas, "Psychosocial Issues in Long-Term Space Flight: An Overview," *Gravitational and Space Biology Bulletin* 12, no. 2 (2001): 25–33.

space medicine in the 1940s.[17] Research to support pilots flying very fast and very high provided a basis for sending astronauts into space. The first conference with "space" in the title was prior to 1950, notes Weitekamp, but some space-oriented research was clandestine or integrated into aviation medicine and psychology in order to avoid the wrath of superiors who thought it wasteful to study "Buck Rogers" issues. In 1961, Bernard Flaherty edited *Psychophysiological Aspects of Space Flight*, which focused on the sensory and biotechnical aspects of spaceflight and simulations, as well as addressing issues of human durability.[18] *Human Factors in Jet and Space Travel* also appeared that year.[19] The latter was edited by Saul B. Sells, a NASA consultant who first wrote about astronaut selection and training in 1957, and Charles A. Berry, at one time NASA Director of Life Science and physician to the astronauts. They dealt with performance under conditions of acceleration and deceleration, as well as human adaptation to space. In 1967, Joseph Kubis, along with Edward J. McLaughlin, specifically addressed the psychological aspects of spaceflight.[20] They noted that whereas short-term spaceflight did not have adverse effects on functioning, factors such as emotional stability and group dynamics could prove important in future missions. As would many other writers, they illustrated their points with studies of psychological reactions to isolation and confinement in terrestrial settings.

In the early 1970s, Joseph Kubis addressed issues of group dynamics: group composition, leadership, and teamwork.[21] In 1971, Air Force psychiatrist Nick Kanas, in collaboration with William E. Fedderson, released an outline of many of the psychological and psychiatric issues that have filtered down and influence discussions today.[22]

---

17. M. A. Weitekamp, *Right Stuff Wrong Sex: America's First Women in Space Program* (Baltimore, MD: Johns Hopkins University Press, 2004).

18. B. E. Flaherty, *Psychophysiological Aspects of Space Flight* (New York: Columbia University Press, 1961).

19. S. B. Sells and C. A. Berry, eds., *Human Factors in Space and Jet Travel: A Medical-Psychological Analysis* (New York: Ronald Press, 1961).

20. J. F. Kubis and E. J. McLaughlin, "Psychological Aspects of Spaceflight," *Transactions of the New York Academy of Sciences* 30, no. 2 (1967): 320–330.

21. J. F. Kubis, "Isolation, Confinement, and Group Dynamics in Long Duration Spaceflight," *Acta Astronautica* 17 (1972): 45–72.

22. N. Kanas and W. E. Fedderson, *Behavioral, Psychiatric, and Sociological Problems of Long Duration Missions* (Washington, DC: NASA Technical Memorandum X-58067, 1971).

Introduction: Psychology and the U.S. Space Program

In 1972, the National Academy of Sciences released the report of a study panel chaired by Donald B. Lindsley of the University of California, Los Angeles (UCLA).[23] The panel sought "to indicate the blocks of research, roughly in order of priority that will be most fruitful in the years ahead in coming to grips with the problems of long-duration missions . . . . In this, there is little doubt in the minds of the study participants that the difficulties are formidable, the unknowns significant, and the prerequisite research extensive . . . ."[24] Many of the experts were interested in space physiology and medicine, but the panel also included psychologists with expertise in stress, social interaction, and behavior in unusual environments. In addition to recommending basic biomedical and life-support research, the panel urged studies of skilled performance, environmental habitability, group processes, interpersonal interaction, and the relationship of the space crew "microsociety" to the larger flight team. In 1977, partially in response to Lindsley's report, Mary M. Connors and her associates began a review of the then-current foundations for understanding behavior during anticipated Space Shuttle and space station missions.[25] Their report, not published until 1985, identified a middle ground between narrowly focused experiments and bold generalizations. They adopted an open systems approach and addressed topics at the individual, small group, and organizational levels.

In the late 1980s, the Committee on Space Biology and Medicine of the National Research Council gave further impetus to psychology, noting that "[a]lthough the evidence is fragmentary, it seems likely that behavioral and social problems have already occurred during long-term missions . . . . An understanding of the problems and their amelioration is essential if man desires to occupy space for extended periods of time. Even more important from a scientific perspective, it seems likely that significant advances in our basic knowledge of human interaction

---

23. D. B. Lindsley, ed., *Human Factors in Long Duration Spaceflight* (Washington, DC: National Academy of Sciences, 1972).

24. Ibid., p. 15.

25. M. M. Connors, A. A. Harrison, and F. R. Akins, *Living Aloft: Human Requirements for Extended Spaceflight* (Washington, DC: NASA SP-483, 1985).

and processes will emerge from the research needed to ensure effective performance and adjustment in space."[26]

Revisiting the issue some 10 years later, a subsequent Committee on Space Biology and Medicine reaffirmed the urgency of their predecessors' recommendations: "Despite this [the 1987 panel's] assessment of the importance of behavioral issues, little progress has been made transforming the recommendations for research on human behavior and performance in space into action . . . . As could be predicted from controlled simulation studies, the history of space exploration has seen many instances of reduced energy levels, mood changes, poor interpersonal relations, faulty decision-making, and lapses in memory and attention. Although these negative psychological reactions have yet to result in disaster, this is no justification for ignoring problems that may have disastrous consequences. Furthermore, there are degrees of failure short of disaster and degrees of success short of perfection; if favorable organizational and environmental conditions can increase the level and probability of success, they are worthy of consideration."[27]

The 1998 Committee's recommendations included studying the effects of the physical and psychosocial environment of spacecraft on cognitive, psychophysiological, and affective measures of behavior and performance; the development and evaluation of countermeasures for mitigating adverse effects of the physical and social environments on individual and group performance; in-flight studies of the characteristics of sleep during long-duration missions; ground-based studies of change and stability in individual psychophysiological patterns in response to psychosocial and environmental stressors; the effects of individual differences on cognitive, psychophysiological, and affective measures of behavior and performance; improved methods for assessing interpersonal relations and crew compatibility; and improved training [didactic and experiential] in psychological and social adaptation to space. The Committee also urged exploring the effects of crew composition on crew tension, cohesion, and performance; factors affecting ground-crew communication and interactions; and conditions that affect the distribution of authority, decision-making, and task assignments between space crews and ground control.

---

26. Committee on Space Biology and Medicine, *A Strategy for Space Medicine and Medical Science for the 1980s and 1990s* (Washington, DC: National Academy Press, 1987), p. 169.

27. Committee on Space Biology and Medicine, *A Strategy for Research in Space Biology and Medicine in the New Century* (Washington, DC: National Academy Press, 1998), p. 169.

Introduction: Psychology and the U.S. Space Program

In 2001, the National Academy of Sciences issued *Safe Passage: Astronaut Care for Exploration Missions*,[28] prepared by the Committee on Creating a Vision for Space Medicine During Travel Beyond Earth Orbit of the Institute of Medicine of the National Academy of Sciences. This panel of experts identified some of the medical and behavioral issues that should be resolved quickly in anticipation of a return to the Moon and a mission to Mars. This far-ranging work covers astronaut health in transit to Earth orbit and beyond, health maintenance, emergency and continuing care, the development of a new infrastructure for space medicine, and medical ethics. Most importantly for present purposes, *Safe Passage* includes a chapter on behavioral health, a topic that we discuss in some detail in chapter 2.

Different missions raise different questions about human behavior. The most conspicuous questions of the earliest days of spaceflight had to do with life support, the human-machine interface, and the optimization of human performance to ensure mission success. Certainly these topics remain crucial today, but to them we may add many more. Following Apollo and the race to the Moon, NASA entered new eras in 1981, when the Space Shuttle took flight, and again in 1993, when astronauts joined cosmonauts first on Russia's *Mir* space station and then on the International Space Station (ISS) in 2000. Topics such as habitability, loneliness, cultural conflicts, the need to sustain a high level of performance over the long haul, and postflight adjustment gained a degree of immediacy and could no longer be ignored. Consistent with Davis Meister's views on conceptual changes in human factors, there has been, over the years, a shift from a purely "displays and knobs" orientation to a more holistic approach, with project managers, engineers, and behavioral researchers sharing the goal of a seamless human-machine structure or "system integration."[29]

In their discussion of post-Apollo psychological issues, Connors and her associates noted that as missions change, so do behavioral requirements.[30] Perhaps the most conspicuous trends are in the direction of increased crew size, diversity, and mission duration. The first round of U.S. flights, under Project Mercury, were solo but rapidly gave way to two-person crews with the advent of Project Gemini in

---

28. J. R. Ball and C. H. Evans, eds., *Safe Passage: Astronaut Care for Exploration Missions* (Washington, DC: National Academy Press, 2001).
29. Meister, *Conceptual Aspects of Human Factors*.
30. Connors et al., *Living Aloft*.

1965, followed by three-person crews during the Apollo program. After Mercury, note Clay Foushee and Robert Helmreich, the test pilot became a less relevant model than the multi-engine aircraft commander, who not only requires technical skills but also requires human relations skills as the leader of a team.[31] America's first space station, *Skylab*, provided a "house in space" for three-person crews; apart from occasional emergencies or visitors, three-person crews were also typical for Soviet (1970–89) and then Russian (1990 and onwards) space stations and the ISS. Space Shuttles are relatively capacious and usually carry six to eight crewmembers. Other than during brief visits from Shuttle crews, the ISS has been home to crews of two to six people. We suspect that later space stations will house larger crews. Although it is possible to envision huge orbiting platforms and communities on the Moon and Mars, foreseeable missions are unlikely to exceed eight people, so crews will remain within the "small group" range.

A second salient trend is toward increasing diversity of crew composition. The initial vision was for a highly diverse pool of astronaut candidates, including mountain climbers, deep sea divers, and arctic explorers, but, as will be explained in the next chapter, it was military test pilots who got the nod. The military remains well represented, but over the years, the astronaut corps has been expanded to include people from many different professions and a greater number of women and minorities. Further complexity was added with the Soviet guest cosmonaut program beginning in the 1970s, the inclusion of international crewmembers on the Shuttle, and international missions on *Mir* and the ISS. Already, tourists have entered the mix, and the first industrial workers in commercial space ventures may not be far behind.

Third, initial spaceflights were measured in hours, then days. (Indeed, within each series of flights, successive Mercury and then Gemini flights were longer and longer, to establish that astronauts could withstand the long trip to the Moon.) The third *Skylab* crew remained on orbit 84 days. *Skylab* was short-lived, but the Soviets set endurance records in this area; the present record of 366 days was set by a Russian cosmonaut on *Mir* during a 1987–88 mission. ISS missions have usually lasted about three months, but individuals are staying on the Space Station for up to six months, as demonstrated in 2007 and 2008 by Sunni Williams and Peggy

---

31. H. C. Foushee and R. L. Helmreich, "Group Interactions and Flight Crew Performance," in *Human Factors in Aviation*, ed. E. L. Wiener and D. C. Nagel (New York: Academic Press, 1998), pp. 189–228.

Introduction: Psychology and the U.S. Space Program

Whitson. Extended stays can also result from unexpected circumstances, such as the loss of the Shuttle *Columbia*, which delayed the retrieval of one crew. If and when humans go to Mars, the sheer distance may require a transit time of two years.

Technology is advancing in all areas, including space exploration. Over the years, electromechanical gauges that dominated cockpits were replaced first with cathode-ray tubes and now with digital displays. New technology is leading to new human-machine partnerships, with computer-based decision aids, improved communications, and increased availability of automated systems and robotics. New on-board systems will augment the astronauts' ability to diagnose and solve flight problems, and it is reasonable to expect improved launch and recovery systems.[32]

In the chapters that follow, psychologists with strong interests in space discuss selected research topics within their historical contexts. In chapter 2, we trace the uneven course of psychology in the space program and describe the history of astronaut selection and psychological support. In chapter 3, Sheryl Bishop points out that whereas there has been limited opportunity to study astronauts in space, there has been ample opportunity to study people in environments that in some ways resemble that of space. These analogs include polar camps and undersea research vessels that share danger, deprivation, isolation, confinement, and other characteristics with spacecraft, along with simulators intended to imitate or mimic spaceflight conditions. In comparison to studies conducted in "everyday" or laboratory settings, studies set in these more extreme environments offer a balance between accessibility and experimental control on the one hand and a degree of environmental realism on the other. Bishop discusses a wide range of analogs and simulators in the United States and abroad and notes that these are absolutely crucial for training purposes.

Spaceflight has positive and rewarding as well as stressful characteristics, and in chapter 4, Julie Robinson, Kelley J. Slack, Valerie Olson, Mike Trenchard, Kim Willis, Pam Baskin, and Jennifer Boyd discuss one of these psychological benefits: observing Earth. They present a unique study of taking pictures from space. This is an excellent example of an unobtrusive study, that is, one that does not set up expectations on the part of the research participants or infringe on their privacy. An overwhelming proportion of the photographs taken from the ISS are initiated

---

32. J. W. McCandless, M. K. Kaiser, T. Barth, R. S. McCann, N. J. Currie, and B. Woolford, "Human-Systems Integration Challenges for Constellation," *Human Factors and Ergonomics Society Annual Meeting Proceedings, Aerospace Systems* 5 (2007): 96–100.

by crewmembers. What kinds of substitute activities can we devise for some future missions when looking out the window may not be an option?

Then, Harvey Wichman points out that soon, spaceflight may no longer be a government monopoly and future spacefarers may include growing proportions of tourists and industrial workers. This situation may require departing from the government agency form of organization that has dominated space exploration so far in favor of a private enterprise model of commercial space exploration; it will also require accommodating people who lack the qualifications of today's astronauts and cosmonauts. In his view, society is at a historical threshold that will require a shift in how engineers, designers, flight managers, and crews perform their tasks. He illustrates some of these points with his industry-sponsored simulation study intended to gauge tourist reactions to spaceflight.

Group dynamics is a focal point for Jason Kring and Megan A. Kaminski, who explore gender effects on social interaction and the determinants of interpersonal cohesion (commitment to membership in the group) and task cohesion (commitment to the work at hand). Their review of the basic literature on mixed-gender groups, as well as findings from spaceflight and other extreme environments, points to the conclusion that whereas there are many benefits to mixed-gender crews (typically, men and women bring different skills to the mix), the issue is multifaceted and complex and poses challenges for spaceflight operations. Although psychologists are gaining some understanding of the determinants of crew cohesion, the effects on performance depend upon the type of cohesion (interpersonal and task) and the nature of the task. None of this is simplified by findings that cohesion is likely to fluctuate over the course of an extended mission.

Cross-cultural issues dominate the next two chapters. In "Flying with Strangers," Peter Suedfeld, Kasia Wilk, and Lindi Cassel draw a distinction between multinational crews, in which "guests" were allowed to participate in U.S. or Soviet/Russian missions, and international crews, which first appeared aboard the International Space Station, which is not owned and operated by any one nation. Through studying the reminiscences of majority and minority participants in multinational and international missions, they test the hypothesis that multinational flights are a source of frustration and annoyance that are not evident in the true partnerships of international flights. Then, Juris Draguns and Albert Harrison elaborate on cross-cultural issues and propose applying a cultural assimilator to build cross-cultural awareness and sensitivity.

Introduction: Psychology and the U.S. Space Program

In the final chapter of this book, Gro Sandal and Gloria Leon present a summary and integration that places the earlier chapters within broader historical, cultural, and organizational contexts. They point out that whereas we can point with pride to past accomplishments, missions will continue to change and there will always be a need for more research and new operational procedures. The research that is done—and, perhaps more importantly, that is not done—reflects political as well as scientific and operational concerns. Many of psychology's advances within the American program are recent, and it is not clear if these gains will withstand the test of time. However, sponsors of other programs, such as the European Space Agency (ESA), understand that psychology is one of the many disciplines required to ensure successful spaceflight.

We conclude our introduction with three important caveats. First, although most of the chapters in this book are authored or coauthored by psychologists and make repeated references to psychology, understanding and managing human behavior in space is an interdisciplinary effort. In essence, "spaceflight psychology" includes contributions from architecture and design, engineering, biology, medicine, anthropology, sociology, communications, and organizational studies, as well as many hybrids (such as cognitive science) and disciplines within psychology (such as environmental, social, and clinical). In a similar vein, the delivery of psychological services to astronauts involves physicians, psychiatrists, social workers, and peers, as well as psychologists.

Second, no one pretends that the chapters in this volume are representative of psychology (never mind the broader field of behavioral science) as a whole. Our essays do not provide in-depth treatment of the interface between engineering and psychology, nor do they attend to the interface of biology and behavior, for example, the effects of cumulative fatigue and circadian rhythms on performance and risk. With respect to this, we note a recent chapter by Barbara Woolford and Frances Mount that described how, over the past 40 years, research on anthropometrics, biomechanics, architecture, and other ergonomics issues slowly shifted from surviving and functioning in microgravity to designing space vehicles and habitats to produce the greatest returns for human knowledge.[33]

---

33. Barbara Woolford and Frances Mount, "Human Space Flight," in *Handbook of Human Factors and Ergonomics*, ed. Gavriel Salvendi, 3rd ed. (Hoboken, NJ: John Wiley and Sons, 2006), pp. 929–955.

Finally, apart from psychological studies of astronauts, we acknowledge many other areas where psychology interfaces with NASA. For example, NASA maintains an excellent program in aviation human factors. Even robotic missions, such as those already dispatched to Mars, have a human touch. It is necessary to assemble, organize, and train teams to manage such missions. Considerable preparation is necessary for successful teleoperations, for example, Earth-bound researchers conducting a "glove box" experiment aboard a satellite thousands of miles away or driving a teleoperated rover on Mars. Some automated satellites are designed for easy servicing by Shuttle crews. Satellites devoted to remote sensing must be designed with human sensory, perceptual, and information processing systems in mind. Furthermore, the loss of *Challenger* and *Columbia* reflected organizational and behavioral factors such as miscommunication and faulty judgment as well as technical failures.[34] Astronauts in flight are the focal point of this volume, but there are many areas where psychology can contribute to NASA.

---

34. M. M. McConnell, *Challenger: A Major Malfunction* (New York: Doubleday, 1987); D. Vaughan, *The Challenger Launch Decision: Risky Technology, Culture and Deviance at NASA* (Chicago: University of Chicago Press, 1996).

Chapter 2
Behavioral Health

Albert A. Harrison
    Department of Psychology
    University of California, Davis

Edna R. Fiedler
    National Space Biomedical Research Institute
    Baylor College of Medicine

ABSTRACT

Experience gained from test pilots, high-altitude balloonists, and animals sent on rocket flights was the starting point for understanding astronaut adaptation and performance in space. Psychology played a significant role in Project Mercury, but before that effort was complete, official interest in such topics as astronaut selection, psychosocial adjustment, group dynamics, and psychological support all but disappeared. Interest was rekindled when astronauts joined cosmonauts on *Mir* and then became full partners on the International Space Station. We review reasons for this period of minimal involvement in the space program and suggest that the "right stuff" image worked against the field until the mid-1990s, when space station expeditions brought the challenges of long-duration missions into focus. Evidence of renewed interest includes the advent of the National Space Biomedical Research Institute, the development of NASA's Bioastronautics Critical Path Roadmap, and the new Human Research Program. In 2001, *Safe Passage: Astronaut Care for Exploration Missions* drew attention to behavioral health, a concept of psychosocial adjustment that depends not only an absence of neuropsychiatric dysfunction but on the presence positive interactions with the physical and social environments. We trace the history and current status of astronaut selection and psychological support, two essential ingredients for maintaining behavioral health, from Mercury to the ISS. Behavioral health is important because it reduces risk, helps optimize performance, and contributes to the welfare of astronauts and their families. We conclude with a brief outline for a comprehensive and continuing program in spaceflight behavioral health.

INTRODUCTION

In the 1950s, as America prepared for its first crewed space missions, it was not clear that human performance capabilities could be maintained under the demanding conditions of spaceflight. Where could NASA begin? Much of the research, equipment, and testing procedures used to support test pilots who set successive speed and altitude records transferred easily to the early space program.[1] Decompression chambers, centrifuges, rocket sleds, and the like made it possible to explore the physiological and performance aspects of conditions that would be encountered in space. Craig Ryan has detailed the contributions of high-altitude ballooning, highlighting the usefulness of gondola designs (which he contends provided a basis for the Mercury spacecraft), flight suits, helmets, and much more.[2] Not everything could be "off the shelf"; NASA had to develop elaborate simulators for upcoming space missions. But, on the whole, the same "cast of characters"—engineers, physicians, and psychologists, to mention a few—who brought America to the edge of space brought America into space.

Animal studies gave some reassurance that humans could adapt physiologically and behaviorally to space.[3] As early as the late 1940s, biological specimens were launched on balloons and sounding rockets. In 1958, the Russians successfully launched a dog, Laika, who survived several days in orbit even though she could not be brought back to Earth. Wernher von Braun approached behavioral biologist Joseph V. Brady to see if he would be willing to launch primates, which would leapfrog the Soviets' dogs.[4] In 1958 and 1959, America's first primate spacefarers, two squirrel monkeys named Able and Baker (known at that time as Miss Able and Miss Baker) were launched on 15-minute flights reaching an altitude of 300 miles on a 1,500-mile trajectory and were successfully recovered following splashdown.

---

1. T. Wolfe, *The Right Stuff* (New York: Farrar, Strauss and Giroux, 1979); M. A. Weitekamp, *Right Stuff, Wrong Sex: America's First Women in Space Program* (Baltimore, MD: Johns Hopkins University Press, 2004).

2. C. Ryan, *The Pre-Astronauts: Manned Ballooning on the Threshold of Space* (Annapolis, MD: Naval Institute Press, 1995).

3. C. Burgess and C. Dubbs, *Animals in Space: From Research Rockets to the Space Shuttle* (Chichester, U.K.: Springer Praxis, 2007).

4. Anon., "Journal Interview 64: Conversation with Joseph V. Brady," *Addiction* 100 (2005): 1805–1812.

One of the main questions was whether the test animals could keep their wits about them in the sense that they could do what they had been taught to do during the presumably terrifying rocket rides. Able and Baker were encased in casts to protect them against gravitational changes, but one finger and one toe were exposed so that, after a warning light turned on, the finger could be used to press a lever to avoid a shock to the toe. All the way up and all the way down, they pressed the lever on cue. Later, as a part of the Mercury pretest program, the chimpanzees Ham and Enos received much more elaborate and sophisticated training than did their predecessors.[5] They flew in special couches within Mercury capsules; Ham's flight was suborbital, but Enos completed four orbits. Although acceleration and deceleration forces in excess of 7 g's had an immediate effect on the chimpanzees' performance, once these forces diminished, their performance bounced back to preflight levels. Microgravity did not interfere with visual processes (monitoring the lights), nor did it interfere with eating and drinking. Not only did they perform their assigned tasks in space, but the two chimpanzees also returned to Earth in good health and with their sharply honed skills intact.[6] Looking back at an episode from this era, Joseph Brady recounted:

> On the recovery ship, after the helicopter had dropped the capsule once or twice before obtaining a good connection on one of these animal pre-test flights—a good reason for practicing before the human flights—the hatch was opened on the flight deck and the chimp came out sputtering and thrashing about. An admiral standing on the deck with several of us said something like "If that chimp could only talk", in response to which I felt required to observe that the best thing that ever happened to us was that the chimp could not talk or the space program might have come to an abrupt end right on the spot.[7]

---

5. F. H. Rholes, Jr., M. E. Grunzke, and H. H. Reynolds, "Chimpanzee Performance During the Ballistic and Orbital Project Mercury Flights," *Journal of Comparative and Physiological Psychology* 86, no. 1 (1963): 2–10.

6. J. V. Brady, "Behavioral Health: The Propaedeutic Requirement," *Aviation, Space, and Environmental Medicine* 76, no. 6, sect. II (June 2005): B13–B24.

7. Anon., " Journal Interview 64": 1811.

During the early 1960s, the United States and Soviet Russia were locked in a race to the Moon, and in many ways, the two programs paralleled each other. In the United States, solo missions (Mercury) gave way to two-person missions (Gemini) and then to three-person missions (Apollo) that, in July of 1969, brought astronauts to the Moon. The Apollo Applications Program followed close on the heels of the last astronaut's departure from the Moon. Based on leftover Moon race equipment, the Apollo Applications Program included the Apollo-Soyuz Test Project, where Americans and Soviets joined spacecraft to live together briefly in space, and *Skylab*, America's "house in space" in the mid-1970s.[8] By the late 1970s, the U.S. and Soviet programs were following different paths: Americans awaited the orbiter, or Space Shuttle, and Soviets launched a series of space stations. In 1984, President Ronald Reagan approved the development of a U.S. space station, but construction was delayed almost 15 years. President Bill Clinton approved the station as a multinational venture, and it became the International Space Station, or ISS. Prior to its construction, American astronauts joined Russian cosmonauts on *Mir*; later, they worked together as partners on the ISS. The ISS recently reached its 10th anniversary of having multinational crews living and working in space.

Although psychology played a prominent role in the early U.S. space program, some branches had all but disappeared by 1963. To be sure, psychologists did show professional interest in humans in space, and many panels and commissions sought to increase psychology's involvement (see chapter 1). Since there were practically no studies of astronauts, researchers relied heavily on studies conducted in Antarctica, submarines and research submersibles, and simulators. Research continues in all three venues; Antarctica took an early lead and remained prominent for many years.[9] A primary reason was that International Geophysical "Year" (IGY, 1957–59) stimulated research on human adaptation to isolation and confinement, with the authoritative and influential accounts appearing in the early 1970s.[10]

---

8. H. S. F. Cooper, Jr., *A House in Space* (New York: Bantam Books, 1976).

9. L. A. Palinkas, "The Psychology of Isolated and Confined Environments: Understanding Human Behavior in Antarctica," *American Psychologist* 58, no. 3 (2003): 353–363.

10. E. K. E. Gunderson, *Human Adaptability to Antarctic Conditions* (Washington, DC: American Geophysical Union, 1973); J. E. Rasmussen, ed., *Man in Isolation and Confinement* (Chicago: Aldine, 1973).

Other factors that favored Antarctica were the large number of people who ventured there and that, as an international site, it offers opportunities for researchers from many different nations. By picking and choosing research locations, one can find conditions that resemble those of many different kinds of space missions, ranging from relatively luxurious space stations to primitive extraterrestrial camps.[11] In 1963, Robert Voas, one of the early space human factors experts, and E. K. Eric Gunderson, who had conducted pioneering psychological research in Antarctica, seriously discussed developing a space mission simulator there, an idea that reemerges from time to time.[12] By the 1980s, it was recognized widely that Antarctica provided a useful meeting ground for people who were interested in adaptation to polar environments and people who were interested in adaptation to space. In 1987, NASA and the National Science Foundation's Division of Polar Programs joined together to sponsor the "Sunnyvale Conference," which brought together researchers from each tradition. Presentations centered on environments (Antarctica and space), theoretical perspectives, isolation and confinement effects, and interventions and outcomes.[13] Antarctic behavioral research became a truly international venture guided in part by the Scientific Committee for Antarctic Research and funded by many sources, including NASA. For example, Des Lugg of NASA Headquarters and Joanna Woods at Johnson Space Center conducted medical and psychological research with the Australian National Antarctic Research Expeditions.[14] The next chapter provides further discussion of analog environments.

---

11. D. T. Andersen, C. P. McKay, R. A. Wharton, Jr., and J. D. Rummel, "An Antarctic Research as a Model for Planetary Exploration," *Journal of the British Interplanetary Society* 43 (1990): 499–504.

12. E. K. E. Gunderson, "Preface," in *From Antarctica to Outer Space: Life in Isolation and Confinement*, ed. A. A. Harrison, Y. A. Clearwater, and C. P. McKay (New York: Springer, 1990), p. 1.

13. A. A. Harrison, Y. A. Clearwater, and C. P. McKay, "The Human Experience in Antarctica: Applications to Life in Space," *Behavioral Science* 34, no. 4 (1989): 253–271; Harrison et al., *From Antarctica to Outer Space*.

14. J. Wood, L. Schmidt, D. Lugg, J. Ayton, T. Phillips, and M. Shepanek, "Life, Survival and Behavioral Health in Small Closed Communities: 10 Years of Studying Small Antarctic Groups," *Aviation, Space, and Environmental Medicine* 76, no. 6, sect. II (June 2005): B89–B94; D. J. Lugg, "Behavioral Health in Antarctica: Implications for Long-Duration Space Missions," *Aviation, Space, and Environmental Medicine* 76, no. 6, sect. II (June 2005): B74–B78.

As noted in chapter 1, despite repetitive calls for action, empirical research was slow to accumulate. In the late 1990s, the National Academy of Sciences undertook a comprehensive review of behavioral and medical issues that we need to begin to address right now to prepare for future space missions. We consider the Academy's report, *Safe Passage: Astronaut Care for Exploration Missions*, a watershed event.[15] Like earlier calls to action, *Safe Passage* drew attention to many biomedical, behavioral, and psychological issues and emphasized their importance for health, performance, and welfare on extended-duration missions. The timing was good because its production and distribution coincided with American missions on board *Mir* and the first missions to the ISS. Although future-oriented, it was developed in the context of unfolding events on then-contemporary extended-duration missions. Most importantly, this work also introduced the concept of behavioral health, an idea that may be particularly useful because of its breadth and relative lack of pejorative connotations. According to one recent definition, "Compared with earlier formulations (such as mental health), behavioral health is less limited in that it recognizes that effective, positive behavior depends on an interaction with the physical and social environments, as well as an absence of neuropsychiatric dysfunction. Behavioral health is evident not only at the level of the individual, but also at the levels of the group and organization."[16]

NASA's recognition of the field of behavioral health and linking of it to performance opened the door for many of the kinds of research that earlier were thought to be too "soft" to be useful to the space program.[17] Today, NASA has shown increased recognition of shared perspectives, privacy, leisure-time activity, family separation and reunification, cultural awareness, the satisfying properties of windows and view ports, and many other topics that were formerly overlooked if not seen as irrelevant or frivolous. From NASA's perspective, the significance of these factors is less in the fact that they can help people "feel good" (although many psy-

---

15. J. R. Ball and C. H. Evans, eds., *Safe Passage: Astronaut Care for Exploration Missions* (Washington, DC: National Academy Press, 2001).

16. A. A. Harrison, "Behavioral Health: Integrating Research and Application in Support of Exploration Missions," *Aviation, Space, and Environmental Medicine* 76, no. 6, sect. II (June 2005): B3.

17. *Bioastronautics Critical Path Roadmap*, http://bioastroroadmap.nasa.gov/index.jsp (accessed 29 March 2008).

chologists would argue that this is a major benefit) than in their potential impact on risk and performance. This research, in turn, has implications for organizing and staging space missions. Thus, a combination of maturing social science and interest sparked by space station and exploration missions has opened the door, at least partially, for new kinds of psychological research within the U.S. space program. Whether this door will remain open—or slam shut—remains to be seen.

THE RIGHT STUFF

For decades, expanding the role of psychology in the U.S. space program was an uphill battle with psychologists' pleas generally falling on deaf ears. Among the more obvious interpretations, it might be tempting to think of NASA managers and engineers as "thing" people rather than "people" people, so the behavioral side of spaceflight is of little interest to them. Perhaps mission managers were simply unaware of the significance of behavioral factors. Or maybe, as "hard" scientists, they saw the behavioral and social sciences as fuzzy and inexact efforts that lead to qualitative recommendations that are difficult to implement and unlikely to work. The sociologist Charles Perrow has discussed how resistance to human factors within complex organizations has strong structural and cultural underpinnings and is not overcome easily.[18]

Psychologists make contributions to human welfare in such diverse areas as environmental design, problem-solving, decision-making, leadership, and group performance, but many people strongly associate psychology with mental illness and long-term psychotherapy. If such attitudes explained NASA's ambivalence about behavioral factors, education would be the antidote; but for many years, educational efforts had little visible impact in research or mission operations.

The stereotype of clinical psychologists and psychiatrists working with troubled clients may have threatening implications for NASA administrators who need to maintain good public relations and build government support. The historian Roger Launius points out that from the moment they were introduced to the public in

---

18. C. E. Perrow, "The Organizational Context of Human Factors Engineering," *Administrative Science Quarterly* 28, no. 4 (1983): 521–541.

1959, America was enthralled by the "virtuous, no nonsense, able and professional astronauts" who "put a very human face on the grandest technological endeavor in history" and "represented the very best that we had to offer."[19] From the beginning, the press was never motivated to dig up dirt on the astronauts; rather, reporters sought confirmation that they embodied America's deepest virtues. "They wanted to demonstrate to their readers that the Mercury seven strode the Earth as latter-day saviors whose purity coupled with noble deeds would purge this land of the evils of communism by besting the Soviet Union on the world stage."[20] Today, people look back longingly to a simpler era when good was good and evil was evil, and, at least in memory, heroes did not disappoint. Psychological research or, worse yet, the faintest possibility that a mission would be compromised by psychological factors could be a public relations nightmare.

For project managers and engineers, faith in the right stuff helps cut costs because the person can be engineered out of the equation. This faith simplifies and speeds the design process as there is no need to waste time consulting behavior experts. Sliding by psychological issues preserves autonomy and decision-making power. If behavioral professionals were to serve in an advisory capacity, mission directors would have to share control, or at least seriously consider the opinion of behavioral experts. Why should managers complicate their task by bringing more players—psychologists, psychiatrists, anthropologists, human factors experts—to the table?

For astronauts, the stereotype of the right stuff helps maintain flight status.[21] It deters snooping and prying that might suggest a real or imagined blemish that could lead to mission disqualification, a most undesirable personal consequence. After all, part of the heroic myth is that under the greatest of adversities, people with the right stuff can still get the job done! Why risk all by getting involved in a research program that could lead to new reasons for disqualification? George Low, manager of Project Apollo, advised subordinates that identity issues, past or present, were

---

19. R. D. Launius, "Heroes in a Vacuum: The Apollo Astronaut as Cultural Icon" (American Institute of Aeronautics and Astronautics [AIAA] Aerospace Sciences Meeting and Exhibit, Reno, NV, 13 January 2005), p. 4.
20. Ibid., p. 4.
21. P. A. Santy, *Choosing the Right Stuff: The Psychological Selection of Astronauts and Cosmonauts* (Westport, CT: Praeger/Greenwood Publishing Group, 1994).

off-limits and that personal hang-ups should be put aside in favor of the mission.[22] Michael Collins and his colleagues liked the John Wayne–type image created for the early astronauts and did not want it tarnished.[23] Flying in space was a macho, masculine endeavor, and there were those who made an effort to reserve the term "astronaut" for men, referring to women who sought to fly in space as "astronautrix," "astro-nettes," "feminauts," and "space girls."[24]

Marc Shepanek points out that today's astronauts are very much aware of the possible effects of stress, boredom, and many other factors on safety, performance, and quality of life in space.[25] He notes that while many of them favor research on these topics, not all stand ready to volunteer as test subjects. The concern is that despite strong assurances of confidentiality, one of the results of their participation could be disqualification. This means that operational psychologists cannot also conduct research: the role of the therapist or consulting organizational psychologist must remain sacrosanct with no hints of dual allegiance to research.[26] Many kinds of workers, including those in the military and law enforcement, worry about breaches of confidentiality that have adverse repercussions on their careers. Worries about a breach of confidentiality are periodically reinforced by officials who release information despite assurances to the contrary.

Efforts to protect the astronauts' image are evident in the cordon that NASA public relations and legal teams establish to prevent outsiders from obtaining potentially damaging information, the micromanagement of astronauts' public appearances, and the great care with which most astronauts comport themselves in public. Even today, there are topics that are considered "too hot" to be included in otherwise comprehensive and informed discussions.

---

22. K. McQuaid, "Race, Gender and Space Exploration: A Chapter in the Social History of the Space Age," *Journal of American Studies* 41, no. 2 (2007): 405–434.

23. Weitekamp, *Right Stuff, Wrong Sex.*

24. Ibid., p. 78.

25. M. Shepanek, "Human Behavioral Research in Space: Quandaries for Research Subjects and Researchers," *Aviation, Space, and Environmental Medicine* 76, no. 6, sect. II (June 2005): B25–B30.

26. C. F. Flynn, "An Operational Approach to Long-Duration Mission Behavioral Health and Performance Factors," *Aviation, Space, and Environmental Medicine* 76, no. 6, sect. II (June 2005): B42–B51.

"The right stuff" is an abstraction or ideal type that living, breathing human astronauts approximate but do not fully attain. By the beginning of the 21st century, cracks began to appear in this image. Researchers had long noted behavioral problems in spaceflightlike environments and worried about what might happen during future space missions. Hints of problems came from the Russian space program, which seemed more attuned to the significance of psychological issues. For Americans, conditions that had been heralded since the 1960s became realities in the 1990s when U.S. astronauts joined Russian cosmonauts on Mir, living and working in space for prolonged periods of time with peers from a very different culture. A few astronauts described some of the behavioral challenges that they encountered in space: maintaining high performance in the face of extreme danger, loneliness, and minor conflicts with other crewmembers.[27] On the debit side of the balance sheet, members of isolated and confined groups frequently report sleep disturbances, somatic complaints (aches, pains, and a constellation of flu-like symptoms sometimes known as the "space crud"), heart palpitations, anxiety, mood swings including mild depression, inconsistent motivation, and performance decrements. Crewmembers sometimes withdraw from one another, get into conflicts with each other, or get into disputes with Mission Control. Eugene Cernan reports that the conflicts between the Apollo 7 crew and Mission Control were so severe that the astronauts never flew again.[28] Both Bryan Burrough and Al Holland have described some of the difficulties that U.S. astronauts experienced on Mir.[29] Burrough writes that Soyuz 21 (1976), Soyuz T-14 (1985), and Soyuz TM-2 (1987) were shortened because of mood, performance, and interpersonal issues. Brian Harvey wrote that psychological factors contributed to the early evacuation of a Salyut 7 crew.[30] U.S. researchers and flight surgeons have acknowledged instances of fear, anxiety, depression, sleep disorders, cognitive changes, somatiza-

---

27. B. Burrough, *Dragonfly: NASA and the Crisis On Board Mir* (New York: Harper Collins, 1998).

28. E. Cernan and D. Davis, *The Last Man on the Moon: Astronaut Eugene Cernan and America's Race to Space* (New York: St. Martin's Press, 1999).

29. Burrough, *Dragonfly*; A. W. Holland, "Psychology of Spaceflight," *Journal of Human Performance in Extreme Environments* 5, no. 1 (2000): 4.

30. B. Harvey, *The New Russian Space Program: From Competition to Cooperation* (Chichester, U.K.: Wiley Praxis, 1996).

tion, impulsive behaviors, social withdrawal, cultural misunderstandings, interpersonal frictions, and anger directed toward Mission Control. After their return, some astronauts reported depression, substance abuse issues, marital discord, and jealousy.[31] Astronauts are highly competent, task-oriented people, who, like other highly functional adults, have the normal ups and downs in their moods and social relationships. And, as in the case of other highly functional adults, these ups and downs can sometimes reduce their effectiveness and relationships.

It is not only the normal ups and downs of the individual astronaut that affect the teams and their work, but also the pressures and occasionally dysfunctional dynamics of the organization and Mission Control. The Mercury astronauts lobbied aggressively to fly as pilots rather than to ride as mere passengers ("Spam in a can") whose spacecraft were controlled from the ground.[32] H. S. F. Cooper wrote a well-publicized account of conflict between the Skylab 4 crew and Mission Control.[33] At the heart of the matter was the overprogramming of the astronauts' time. As psychologist Karl Weick described the situation:

> To get the most information from this final trip in the Apollo program, ground control in Houston had removed virtually all the slack from the astronauts' schedule of activities and had treated the men as if they were robots. To get everything in, ground control shortened meal times, reduced setup times for experiments, and made no allowance for the fact that previous crews aboard Skylab had stowed equipment in an unsystematic manner. The astronauts' favorite pastimes—watching the sun and earth—were forbidden.[34]

---

31. Flynn, "An Operational Approach"; Shepanek, "Human Behavioral Research in Space"; P. Suedfeld, "Invulnerability, Coping, Salutogenesis, Integration: The Four Phases of Space Psychology," *Aviation, Space, and Environmental Medicine* 76, no. 6, sect. II (June 2005): B3–B12.
32. Wolfe, *The Right Stuff*.
33. Cooper, *A House in Space*.
34. K. E. Weick, "Organizational Design: Organizations as Self-Designing Systems," *Organizational Dynamics* (Autumn 1977): 31.

Thus, on 27 December 1973, the Skylab 4 astronauts conducted a daylong "sit-down strike." Cooper described the crew pejoratively as hostile, irritable, and downright grumpy, while other writers have described the "strike" as a legitimate reaction to overwork.[35] William K. Douglas, a NASA flight surgeon, lamented both Cooper's emotionally toned reporting and people's willingness to focus on others' real or imagined failures while overlooking greatness.[36] Whatever the "spin" on this particular event, the lessons are clear: the same rapid pace that can be sustained for brief sprints cannot be sustained for marathons. Give astronauts the flexibility to schedule their own activities, and allow time to look out the windows. NASA appears to have taken the lesson to heart. In 2002, Space.com's Todd Halvorson conducted an interview with enthusiastic ISS astronaut Susan Helms. "It's not that the crew isn't busy maintaining the station, testing the remote manipulator and conducting science, it's that there remains enough time to look out the window, do somersaults in weightlessness, watch movies, and sit around chatting."[37]

Spaceflight also offers opportunities for psychological growth and development.[38] Training for and working in space allows people to develop their abilities, gain a strong sense of accomplishment, and feel worthwhile. There is unparalleled challenge, the opportunity to redefine one's place in the cosmos. There is the exhilarating feeling, as Harrison Schmitt wrote, of actually "being there."[39] Walter Cunningham wrote, "It has caused me to seek a challenge wherever I can find one, to charge ahead and never look back . . . that feeling of omnipotence is worth all that it takes to get there."[40] Many of the two dozen or so astronauts and cosmo-

---

35. M. M. Connors, A. A. Harrison, and F. R. Akins, *Living Aloft: Human Requirements for Extended Spaceflight* (Washington, DC: NASA SP-483, 1985).

36. William K. Douglas, "Psychological and Sociological Aspects of Manned Spaceflight," in *From Antarctica to Outer Space*, ed. Harrison et al., pp. 81–88.

37. T. Halvorson, "ISS Astronaut Susan Helms: Space Is More Than a Nice Place to Visit," available at *http://www.space.com/missionlaunches/missions/iss_freetime_010615.html*, 15 June 2001 (accessed 23 June 2010).

38. A. A. Harrison and J. E. Summit, "How Third Force Psychology Might View Humans in Space," *Space Power* 10 (1991): 85–203.

39. H. Schmidt, "The Millennium Project," in *Strategies for Mars: A Guide for Human Exploration*, ed. C. Stoker and C. Emmart (San Diego: American Astronautical Society/Univelt, 1996), p. 37.

40. W. Cunningham, *The All-American Boys* (New York: Macmillan, 1977), p. 27.

nauts interviewed by Frank White reported "overview effects," truly transformative experiences including senses of wonder and awe, unity with nature, transcendence, and universal brotherhood.[41] More recent testimonials concerning the psychological benefits of life in space come from Apollo 14 astronaut Edgar Mitchell and Shuttle-*Mir* astronaut-cosmonaut Jerry Linenger.[42] Astronauts and cosmonauts like the sense of adventure, camaraderie, and grandeur in space.[43]

We find hints of long-term physical and mental health benefits to life in challenging environments. For example, a long-term followup study of Navy personnel who had wintered in Antarctica revealed that following their return, they had undergone fewer hospitalizations than their peers who had identical qualifications but whose orders to go to the South Pole were rescinded as the result of an arbitrary administrative decision.[44] Studies of the mental health of cosmonauts conducted two or three years after their return to Earth found that they had become less anxious, hypochondriacal, depressive, and aggressive.[45] The most plausible explanation is that during their stay in tough environments, people develop coping skills, that is, ways of dealing with challenge and stress that continue to serve them well long after they have returned from their expedition.

It was about the time astronauts began traveling on *Mir* and the ISS that greater evidence of psychology began to show in the U.S. space program. NASA's Bioastronautics Critical Path Roadmap (BCPR) is one piece of evidence. Bioastronautics was NASA's shorthand for life in space, and the BCPR was a framework for identifying the knowledge that NASA needs for future space missions.[46] It identified and assigned priorities to the biomedical and behavioral questions that must be addressed (and the kinds of countermeasures that must be designed) for

---

41. F. White, *The Overview Effect* (Boston: Houghton Mifflin, 1987).

42. E. Mitchell and D. Williams, *The Way of the Explorer* (New York: Putnam, 1996); J. M. Linenger, *Off the Planet* (New York: McGraw-Hill, 2000).

43. Suedfeld, "Invulnerability, Coping, Salutogenesis, Integration."

44. L. A. Palinkas, "Group Adaptation and Individual Adjustment in Antarctica: A Summary of Recent Research," in *From Antarctica to Outer Space*, ed. Harrison et al., pp. 239–252.

45. V. I. Myasnikov and I. S. Zamaletdinov, "Psychological States and Group Interaction of Crew Members in Flight," in *Humans in Spaceflight*, ed. C. L. Huntoon, V. Antipov, and A. I. Grigoriev, vol. 3, bk. 2 (Reston, VA: AIAA, 1996), pp. 419–431.

46. *Bioastronautics Critical Path Roadmap*, available at *http://bioastroroadmap.nasa.gov/index.jsp* (accessed 30 March 2008).

Space Station, lunar, and Mars missions. The BCPR represented a major investment of time and energy, of soliciting and responding to expert advice, and of building consensus. It recognized that NASA's organizational chart was not isomorphic with the way that research is traditionally organized and tried assiduously to address crucial gaps. The BCPR was a useful mechanism for organizing biomedical and behavioral research and fostered research that yielded operationally relevant results. Most importantly, it represented a higher level of "buy-in" to behavioral research on the part of the space agency. Recently, the BCPR has evolved into the Human Research Program. As of January 2010, six elements compose the Human Research Program. They are the International Space Station Medical Project, Space Radiation, Human Health Countermeasures, Exploration Medical Capability, Behavioral Health and Performance, and Space Human Factors and Habitability.[47] As the mission of NASA changes, the exact delineation of the Human Research Program may also change.

Also coincident with turn-of-the-millennium space station missions was the initiation of the National Space Biomedical Research Institute (NSBRI), a consortium of universities and businesses dedicated to solving the problems of astronauts who are undertaking long-duration missions. The NSBRI is best viewed as tightly networked centers of excellence. Members of affiliated organizations form interdisciplinary teams that cut across organizational boundaries and draw strength from one another. The Institute also provides workshops and retreats for investigators who are working under the NSBRI umbrella.

Many of the research interests represented in the NSBRI are clearly biomedical—for example, bone and muscle loss, immune disorders, and radiation effects. Other teams include neurobehavioral and psychosocial factors and human performance. For instance, there are studies of crew composition, structure, communication, and leadership style. Also, there is research on methods to prevent sleep loss, promote wakefulness, reduce human error, and optimize mental and physical performance during long-duration spaceflight. Whereas many organizations hope to extrapolate studies of Earth-bound populations to astronauts and cosmonauts,

---

47. *Human Research Program Evidence Book*, available at *http://humanresearch.jsc.nasa.gov/elements/smo/hrp_evidence_book.asp* (accessed 23 June 2010); NASA, *Human Research Program*, available at *http://humanresearch.jsc.nasa.gov/about.asp* (accessed 25 March 2008).

NSBRI partners hope that their research on spacefarers and analogs will benefit people on Earth.

In 2003, NASA commissioned a workshop on spaceflight behavioral health. The primary purpose of this workshop was to bring together researchers and practitioners in an effort to identify research gaps and produce an archival record for use by managers, established behavioral health researchers, and newcomers to the field.[48] Also, and perhaps most important since the mid-1990s, astronauts have begun to respond to questionnaires on such topics as noise levels and communication.[49] Astronauts have taken part in flight studies involving sleep and circadian rhythms and have taken self-administered tests of cognitive ability, maintained diaries, and provided other information from orbit.[50] Compared to those of earlier years, many of today's astronauts are more willing to participate in ground-based and in-flight studies, given proper assurances of confidentiality.

We suggest that the NASA-*Mir* missions opened a window of opportunity for fruitful reevaluation of the role of behavior, including psychosocial adaptation, in U.S. space missions. When extended-duration missions moved from the abstract and theoretical to the real and some astronauts broached topics like risk, loneliness, and culture conflicts, psychological factors were brought into sharp focus. In policy studies, a window of opportunity opens when a major, unexpected catastrophe (known as a focusing event) becomes known to policy-makers and the public at the same time.[51] Certainly, minor problems on *Mir* were far removed from catastrophic,

---

48. A. A. Harrison, "New Directions in Spaceflight Behavioral Health: A Workshop Integrating Research and Application," *Aviation, Space, and Environmental Medicine* 76, no. 6, sect. II (June 2005): B3–B12.

49. A. D. Kelly and N. Kanas, "Crewmember Communications in Space: A Survey of Astronauts and Cosmonauts," *Aviation, Space, and Environmental Medicine* 63 (1992): 721–726; A. D. Kelly and N. Kanas, "Leisure Time Activities in Space: A Survey of Astronauts and Cosmonauts," *Acta Astronautica* 32 (1993): 451–457; A. D. Kelly and N. Kanas, "Communication Between Space Crews and Ground Personnel: A Survey of Astronauts and Cosmonauts," *Aviation, Space, and Environmental Medicine* 64 (1993): 795–800.

50. M. M. Mallis and C. W. DeRoshia, "Circadian Rhythms, Sleep, and Performance in Space," *Aviation, Space, and Environmental Medicine* 76, no. 6, sect. II (June 2005): B94–B107; R. L. Kane, P. Short, W. E. Sipes, and C. F. Flynn, "Development and Validation of the Spaceflight Cognitive Assessment Tool for Windows (WinSCAT)," *Aviation, Space, and Environmental Medicine* 76, no. 6, sect. II (June 2005): B183–B191.

51. T. A. Birkland, "Focusing Events, Mobilization, and Agenda Setting," *Journal of Public Policy* 18, no. 1 (1997): 53–74.

but behavioral issues gained salience and became known to NASA officials and the public at the same time. The astronauts' experiences on *Mir* opened a window that generated interest in spaceflight behavioral health.

In 1984, Robert Helmreich pointed out that in contrast to Americans, the Russians seemed to have always maintained a certain degree of interest in psychosocial adaptation.[52] He reprinted several quotes from cosmonauts showing interest in psychosocial adjustment, group dynamics, and related topics, and he pointed to the publication of a collection of papers on space psychology by Petrov, Lomov, and Samsonov.[53] Nick Kanas and his associates have written extensively on the role of psychology in the Soviet and then Russian space programs and have highlighted the potential value of this research for NASA.[54] By the mid-1980s, Oleg Gazenko, head of Soviet space medicine, concluded that the limitations of living in space are not medical, but psychological.[55] Quotes from cosmonaut diaries and Soviet/Russian reports remain popular for illustrating the importance of stress, mental health, crew dynamics, and the like, in part because for a long time, neither NASA support personnel nor astronauts themselves freely commented on such issues.

In the early 1970s, there were only three crewed missions, and then America's "House in Space," *Skylab*, was abandoned. The United States invested in the Shuttle, which supports fairly large crews, but for only short times in space. America expected a space station, but it was not approved until 1984, and the station itself underwent several iterations (Space Station, Space Station *Alpha*, and Space Station *Freedom*) before becoming the ISS. The Soviets, on the other hand, moved directly into the era of Salyut and *Mir* space station missions. For them, extended-duration missions—and focusing events in the area of behavioral health—became

---

52. R. L. Helmreich, "Applying Psychology to Outer Space: Unfulfilled Promises Revisited," *American Psychologist* 38 (1983): 445–450; Santy, *Choosing the Right Stuff*.

53. B. N. Petrov, B. F. Lomov, and N. D. Samsonov, eds., *Psychological Problems of Spaceflight* (Moscow: Nauka Press, 1979).

54. N. Kanas, "Psychosocial Factors Affecting Simulated and Actual Space Missions," *Aviation, Space, and Environmental Medicine* 56, no. 8 (August 1985): 806–811; N. Kanas, "Psychosocial Support for Cosmonauts," *Aviation, Space, and Environmental Medicine* 62, no. 4 (August 1991): 353–355; N. Kanas, V. P. Salnitskiy, J. B. Ritsher, V. I. Gushin, D. S. Weiss, S. A. Saylor, O. P. Kozerenko, and C. R. Marmar, "Human Interactions in Space: ISS vs. Shuttle/Mir," *Acta Astronautica* 59 (2006): 413–419.

55. J. E. Oberg and A. R. Oberg, *Pioneering the Space Frontier* (New York: McGraw-Hill, 1986).

a reality decades ago. As Connors and her associates wrote in 1986, "The Russians have experienced longer spaceflights than their American counterparts and have given considerable attention to ways of maintaining individuals' psychological health and high morale in space . . . . In the Soviet Union, the Group for Psychological Support is an acknowledged and welcomed component of the ground team. Concern over such issues as intragroup compatibility and the effects of boredom on productivity seem to be actively studied by cosmonauts and psychologists alike. There appears to be little if any loss of status associated with confirmation of psychological or social problems associated with confinement in space."[56]

Thus, Russians had to confront in the 1970s issues that became pressing for Americans two decades later. As a result, when looking for models for a psychological support program, NASA turned to the Russian program to support cosmonauts on *Mir*.[57] It is interesting that America's international partners in space—European as well as Japanese—share the Russians' interest in spaceflight psychology.[58]

ASTRONAUT SELECTION

NASA, chartered as a civilian space agency, initially intended to select Mercury astronauts from a relatively broad range of explorers: military and commercial aviators; mountain climbers; polar explorers; bathysphere operators; and other fit, intelligent, highly motivated individuals who had demonstrated capabilities for venturing into dangerous new areas. Strong pressure from the White House limited the pool to military test pilots.[59] This was a group of accomplished fliers, many of whom had braved death during war. They brought with them the sharp wits, relentless motivation, and strong emotional control that characterize pilots who are willing to push themselves and their aircraft to (and sometimes beyond)

---

56. M. M. Connors, A. A. Harrison, and F. R. Akins, "Psychology in the Resurgent Space Program," *American Psychologist* 41, no. 8 (August 1986): 906–913.

57. W. E. Sipes and S. T. Vander Ark, "Operational Behavioral Health and Performance Resources for International Space Station Crews and Families," *Aviation, Space, and Environmental Medicine* 76, no. 6, sect. II (June 2005): B3.

58. Santy, *Choosing the Right Stuff*.

59. Ibid.

the limits. Furthermore, because they were under military command, they were used to taking orders and were already cleared for top-secret technology. Mercury candidates had to be under 40 years of age, have graduated from college with a bachelor's degree in science or engineering, have logged at least 1,500 hours flying jet planes, and have graduated from test pilot school. Of course, they were expected to be free of disease or illness and to demonstrate resistance to the physical stressors of spaceflight, such as temperature extremes and rapid acceleration and deceleration. To fit in the cramped confines of the Mercury capsule, their height could not exceed 5 feet 9 inches. The first astronauts had five duties: survive, perform effectively, add reliability to the automated system, complement instrument and satellite observation with scientific human observation, and improve the flight system through human engineering capabilities.[60]

The initial Mercury project used two psychological approaches to selection. One was the industrial-organizational model of select-in characteristics emphasizing astronaut proficiencies needed to successfully complete mission tasks. The second was the psychiatric-clinical psychology model of select-out characteristics. As Robert Voas and Raymond Zedekar point out, psychological qualifications fell into two categories: abilities and personality.[61] In terms of aptitude and ability, they include high intelligence, general scientific knowledge and research skills, a good understanding of engineering, knowledge of operational procedures for aircraft and missiles, and psychomotor skills such as those used to operate aircraft. As regards personality, astronauts were to demonstrate a strong motivation to participate in the program, high tolerance for stress, good decision-making skills, emotional maturity, and the ability to work with others.

At that time, of 508 military test pilots, 110 met the general requirements and 69 were considered highly qualified. These were invited to the Pentagon for a briefing and interviews. Then, 32 were sent to the Lovelace clinic for an extraordinary physical exam and, after certification at Lovelace, to Wright Air Development Center in Dayton, Ohio, for tests of performance under stress. Here, the candidates were subjected to vibration, acceleration and deceleration, sitting with their feet

---

60. M. M. Link, *Space Medicine in Project Mercury* (Washington, DC: NASA, 1965).
61. R. Voas and R. Zedekar, "Astronaut Selection and Training," chap. 10 in *Mercury Project Summary Including the Results of the Fourth Manned Orbital Flight, May 15 and 16, 1963* (Washington, DC: Office of Scientific and Technical Information, NASA, October 1963).

in tubs of ice water, and numerous psychological and psychiatric evaluations. They completed 13 tests on personality and motivation, and another dozen or so on intelligence and aptitudes. NASA historians offer the following observation:

> Two of the more interesting personality and motivation studies seemed like parlor games at first, until it became evident how profound an exercise in Socratic introspection was implied by conscientious answers to the test questions "Who am I" and "Whom would you assign to the mission if you could not go yourself?" . . . . Candidates who proceeded this far in the selection process all agreed with the one who complained "Nothing is sacred any more."[62]

After five Mercury flights, NASA officials decided that, given the absence of serious performance deficits to date, there was no need to continue exhaustive testing procedures. Although ongoing research would have provided an excellent basis for refining selection methods, by the end of 1962, NASA had prohibited research teams from collecting data on astronaut job performance, thus making it impossible to validate selection methods. At that point, according to Patricia Santy's authoritative work, *Choosing the Right Stuff: The Psychological Assessment of Astronauts and Cosmonauts*, normal reluctance to participate in psychological research was transformed into "outright hostility."[63] Psychiatric and psychological data from the Mercury program were confiscated, and researchers were told that apart from incomplete information that had already appeared in an obscure interim report, nothing could be published about astronaut psychology. The reasons for this are not entirely clear—for example, confidentiality was a growing concern, and data that could provide a basis for invidious comparisons could work against crew morale—but Santy favors the view that "NASA became fearful that information on the psychological status and performance of their astronauts would be detrimental to the agency."[64]

---

62. Mercury Program Overview, available at *http://www-pao.ksc.nasa.gov/history/mercury/mercury-overview.htm* (accessed 4 December 2007).
63. Santy, *Choosing the Right Stuff*, p. 29.
64. Ibid., p. 29.

She also documents the minimal role that psychiatrists and psychologists played in the selection process from Gemini until well into the early Shuttle missions.[65] In the beginning of the astronaut program, original psychological selection attempted to pick the best-qualified candidates from a very capable group of experienced pilots, but by the 1980s, the selection process simply made sure that candidates were qualified based on the evaluator's opinion. Thus in 1983, Jones and Annes could claim that no psychological testing was involved. Rather, the approach had evolved into an entirely psychiatric process completed by two psychiatrists who separately interviewed each candidate. Whereas the original examination sought the best-qualified candidates, later procedures simply ensured that each candidate met the minimum qualifications.[66]

Candidates were no longer rated against one another, but they were screened for various psychopathologic conditions that could be detrimental or unsafe in a space environment. This screening, although conducted by expert aviation psychiatrists, did not have specific and objective criteria by which to rate each candidate. The emphasis was on selecting-out those candidates whose psychological structure would be detrimental in a space environment. Neuroses, personality disorders, fear of flying, disabling phobias, substance abuse, the use of psychotropic medications, or any other psychiatric conditions that would be hazardous to flight safety or mission accomplishment were among the grounds for rejection.

Thus, a selection program that began in 1959 as a model rooted in psychiatry and clinical psychology, and in industrial and organizational psychology, had been reduced to subjective evaluation. Patricia Santy provides more detail on how psychiatric evaluations were conducted by two psychiatric consultants who did not collaborate, use a standardized psychiatric interview, or keep detailed documentation, and who used their own subjective sets of psychological criteria in the course of the evaluation.[67] She reviewed the percentage of female and male candidates disqualified psychiatrically. She found that one of the two psychiatrists hired to help in the screening process between 1977 and 1985 psychiatrically disqualified 40.7

---

65. Ibid.
66. D. R. Jones and C. A. Annes, "The Evolution and Present Status of Mental Health Standards for Selection of USAF Candidates for Space Missions," *Aviation, Space, and Environmental Medicine* 54 (1983): 730–734.
67. Santy, *Choosing the Right Stuff*.

percent of the female candidates and 7.5 percent of the male candidates. However, since no specific documentation existed, there was no way to know the reasoning behind his decisions.[68] This is not to say that the psychiatric consultants did a poor job of selecting-out; because no validation studies were completed, there is no evidence by which to evaluate their work.

Under the leadership of psychiatrist Patricia Santy and psychologist Al Holland in the 1980s, and then, in the 1990s, psychiatrist Christopher Flynn, there was a gradual return to evidence- and normative-based astronaut selection. In 1988, a biobehavioral research laboratory was formed within the Space Biomedical Research Institute (SBRI), which at that time was a branch of NASA's Medical Sciences Division, along with Medical Operations. Michael Bungo headed SBRI; Patricia Santy was the director of the laboratory; and psychologist Al Holland became her deputy. The Biobehavioral Laboratory was to develop a new working group of psychologists and psychiatrists to make recommendations on both the operational and research needs in the areas of the behavioral sciences. At that time, operations were expanding beyond helping to choose astronauts to providing psychological support for the astronaut corps.

The development of standardized, semistructured interviews and diagnostic criteria, aided by the work done by the Working Group on Psychiatric and Psychological Selection of Astronauts, resulted in a rewrite of NASA psychiatric standards based on the then-current American Psychiatric Association's *Diagnostic and Statistical Manual III* and recommendations for a select-in process.[69] The reasoning behind the select-in process harkened back to the original logic of 1959, hypothesizing that certain psychological traits were associated with effective astronaut performance. Commencing in 1989, validation work on the select-in criteria was begun. In describing the selection process, Laura Galarza and Al Holland note that selection starts at the time of entry into the astronaut corps, then should continue through the training process and include selection for designated missions.[70]

---

68. Ibid.

69. Ibid.

70. L. Galarza and A. W. Holland, "Selecting Astronauts for Long-Duration Missions" (SAE International Document 1999-01-2097, presented at the International Conference on Environmental Systems, Denver, CO, July 1999); L. Galarza and A. W. Holland, "Critical Astronaut Proficiencies Required for Long-Duration Space Flight" (SAE International

In the 1990s, Galarza and Holland began developing a scientifically defensible select-in process that would screen for personal abilities to help people live and work within small teams under conditions of isolation and confinement.[71] By using highly qualified subject-matter experts, job analysis, and documented validation techniques, they sought to meet the high standards for selection established by the Society for Industrial and Organizational Psychologists (SIOP).[72] Although these researchers developed a profile of needed knowledge, skills, and abilities, NASA's prohibition against obtaining in-training or on-the-job performance ratings effectively killed any longitudinal or predictive validation of the proposed astronaut select-in procedures. Today, all astronaut candidate applicants spend several hours completing psychological tests and then undergo extensive psychological and psychiatric interviews. To prevent coaching, the specific tests and interview content are not publicly available. The current selection process resembles the selection procedures for other high-risk jobs and incorporates highly validated tests that are quantitatively scored, along with in-depth, semistructured interviews.

Well before Apollo astronauts set foot on the Moon, there were political pressures to increase the diversity of the astronaut corps by including women and representatives of different racial and ethnic groups. Accommodating people with different cultural backgrounds became a practical matter in the Apollo-Soyuz rendezvous, in the course of the Russian "guest cosmonaut" program, in Shuttle missions with international crews, and, of course, aboard the ISS. Successfully managing cultural, occupational, and other differences in space is likely to become even more crucial as highly trained professionals are joined by industrial workers and tourists.

Margaret Weitekamp recounts how, at the inception of Project Mercury, an Air Force flight surgeon, Don Flickenger, helped initiate a program known as WISE—Women in Space Earliest.[73] Women offered certain potential advantages over men; one of the most notable of these was their smaller size (and reduced life-support requirements), which would make them easier to lift into orbit and keep alive at

---

Document 1999-01-2096, presented at the International Conference on Environmental Systems, Denver, CO, July 1999).

71. Ibid.

72. Society of Industrial and Organizational Psychology, *Principles for the Validation and Use of Personnel Selection Procedures* (Washington, DC: SIOP, 2003).

73. Weitekamp, *Right Stuff, Wrong Sex.*

a time when engineers had to fret every extra pound of weight. After word of the program's existence leaked, it was abandoned by the Air Force and taken over by Dr. Randall Lovelace, of the same Lovelace Clinic that conducted the physicals for project Mercury. Aviatrix Jackie Cochran and her wealthy philanthropist husband, Floyd Odlum, provided funding so that Lovelace could put the women through the same rigorous evaluation. Of the 25 women who took the physical, 13 passed. The next step in the process, which involved centrifuges and jet flights, depended on the availability of military facilities and equipment. Although it appeared that the procedures could be done at the Naval Air Station in Pensacola, Florida, the ability to do so depended on NASA's officially "requiring" and then reimbursing the testing. Since the program was unofficial (despite widespread perceptions that it was connected with NASA), the space agency did not intervene on the women's behalf. Some of the women continued to press for further testing and flight training, and, eventually, there was a congressional hearing, but public clamor and aggressive lobbying got no results. Kennedy's decision to place a man on the Moon before the decade was finished was interpreted by NASA to mean that it could not divert resources to sending women to orbit. But there were other barriers to women's participation in space exploration, including the inability of some of the people in NASA's white-male-dominated culture to conceive of women in the "masculine" role of astronaut. Weitekamp writes:

> At a very basic level, it never occurred to American decision makers to seriously consider a woman astronaut. In the late 1950s and early 1960s, NASA officials and other American space policy makers remained unconscious of the way their calculations implicitly incorporated postwar beliefs about men's and women's roles. Within the civilian space agency, the macho ethos of test piloting and military aviation remained intact. The tacit acceptance that military jet test pilots sometimes drank too much (and often drove too fast) complemented the expectation that women wore gloves and high heels—and did not fly spaceships.[74]

---

74. Ibid., p. 3.

At that time, lack of diversity at NASA was not limited to the astronaut corps. In 1974, Congress held a hearing on NASA's Equal Employment Opportunity Program. The chairman's introductory remarks included the statement "It is clear that the NASA equal employment opportunity effort over the years has been inadequate . . . ."[75] In the congressional report, NASA admitted that as of the end of fiscal year (FY) 1971, of all NASA employees, only 16.6 percent were women and 4.6 percent minorities.[76] Only 3 percent of the supervisors and 2.4 percent of the engineers were women.

Kim McQuaid points out that many forces worked against increasing the proportion of women and blacks at NASA.[77] Nationally, efforts to increase diversity through new employment strategies began at about the same time as NASA flourished in the late 1960s and early 1970s. Special hurdles at NASA included an organizational culture that was built on the white-male stereotypes of the time and demanded prior training and experience in science and engineering at a time when very few women or minorities were earning (or were allowed to earn) degrees in science and engineering. In 1973, then–NASA Administrator James Fletcher hired Ruth Bates Harris as a high-level deputy director to oversee NASA's equal opportunity employment processes—but, when it turned out that she would be a fearless leader rather than a compliant bureaucrat, he fired her and then, under pressure, attempted to rehire her at a lower level. This initiated bad press, conflicts with Congress, and a series of internal struggles that brought about diversification. In the 1990s, Administrator Dan Goldin could complain that NASA was still too male, pale, and stale, although, two decades earlier, NASA had responded to new domestic political issues by changing from a civil rights sham to the beginnings of a demonstrably effective, if imperfect, affirmative action program.

Aside from the 1965 selection cycle, when the National Academy of Sciences handled selection and allowed women to apply (none were accepted), it was not until the Shuttle era that women were added to the astronaut corps. On 16 January

---

75. House Committee on the Judiciary, Subcommittee on Civil Rights and Constitutional Rights, *NASA's Equal Opportunity Program*, hearings before the Subcommittee on the Judiciary, 93rd Congress, 2nd session, 13–14 March 1974, p. 1.

76. Ibid., p. 13.

77. Kim McQuaid, "Race, Gender and Space Exploration: A Chapter in the Social History of the Space Age," *Journal of American Studies* 41, no. 2 (2007): 405–434.

1978, the first female and black candidates were selected; only a few years later, in 1983, the public wildly acclaimed mission specialist Sally Ride's orbital flight aboard *Challenger*. Some of the women who had participated in the informal women's astronaut selection program of the early 1960s felt vindicated in 1995, when they watched pilot Eileen Collins lift off, carrying their dreams with her.[78] Today, female astronauts routinely participate in Shuttle and Space Station missions in many different roles. Despite the long road that American women and minorities traveled to prove their worth, the U.S. experience has shown that talented women and minorities, given no special treatment because of gender or ethnicity, are as adept as their white, male colleagues in the world of space.

PSYCHOLOGICAL SUPPORT

Initially, psychological support for astronauts came from helpful flight surgeons, flak-catchers who tried to minimize interference on the part of the media and the public, as well as cheering family and friends. By means of shortwave radio, astronauts on the ground encouraged astronauts in orbit. It is clear from Wolfe's *The Right Stuff* that the astronauts' wives provided strong support for one another, as well as for their husbands.[79] The larger community of astronauts and their families still provides psychological support for astronauts before, during, and after their flights.

Professional psychological support for the astronauts and their families evolved over time and gained momentum in the early space station era.[80] Today, psychological support is provided in three stages: preflight, in-flight, and postflight.[81]

---

78. Weitekamp, *Right Stuff, Wrong Sex*, p. 188.
79. Wolfe, *The Right Stuff*.
80. E. Fiedler and F. E. Carpenter, "Evolution of the Behavioral Health Sciences Branch of the Space Medicine and Health Care Systems at the Johnson Space Center," *Aviation, Space, and Environmental Medicine* 76, no. 6, sect. II (June 2005): B31–B35; Flynn, "An Operational Approach to Long-Duration Mission Behavioral Health and Performance Factors"; N. Kanas and D. Manzey, *Space Psychology and Psychiatry* (Dordrecht, Netherlands: Kluwer, 2003).
81. W. E. Sipes and E. Fiedler, "Current Psychological Support for US Astronauts on the International Space Station" (paper presented at "Tools for Psychological Support During Exploration Missions to Mars and Moon," European Space Research and Technology Centre [ESTEC], Noordwijk, Netherlands, 26 March 2007).

The NASA and Wyle Operational Psychology team, under the leadership of the Behavioral Health and Performance Group/Space Medicine, NASA, offers preflight training and briefings in such diverse areas as self-care, conflict management and cultural awareness, and field training. Family readiness is addressed in a briefing focused on the astronaut's spouse to explain processes such as crew care packages and private family conferences. Crew care packages are containers of personal items from family and friends that are sent via Russian Soyuz supply missions and U.S. Space Shuttle missions to astronauts residing on the ISS. Favorite foods, surprise gifts from the family, and holiday decorations are a few of the items that have been sent to the ISS in these shipments.

During the flight stage, in addition to the crew care packages and private weekly videoconferences with families, psychological support services include extensive communication with people on the ground (including Mission Control personnel, relatives, and friends), psychological support hardware and software, special events such as surprise calls from celebrities, and semimonthly videos with a behavioral health clinician. Astronauts in flight have e-mail accessibility and can use an Internet protocol phone on board the ISS to call back to Earth. As in the past, ham radio allows contact between the ISS and schools throughout the world.

A month before their return to Earth, ISS astronauts are briefed on the stresses and joys of returning home following the deployment. Postflight, there are a series of debriefings intended to benefit the astronaut and fine-tune the psychological support program. The astronaut's spouse is given the opportunity to meet with operational psychological support personnel to provide the latter with feedback on the psychological support provided during the mission. Of course, astronauts and their families can use counseling psychological support services at any time. While this briefly covers the current state of the art of psychological support for astronauts on the ISS, psychological support for lunar and Mars missions may have greater constraints and force a return to the mindset of earlier explorers and their families.

CONCLUSION

Spaceflight is both demanding and rewarding, and for many years, psychologists focused on the demanding environment and stressful effects. Throughout the history of spaceflight, psychologists, psychiatrists, and many other professionals have expressed concern that the physical, psychological, and interpersonal stressors of spaceflight could endanger a crew, undercut performance, and lower the quality of life. Episodes in spaceflight-analogous environments and a few incidents in space suggest that although no astronauts have been recalled to Earth on the basis of psychological and social challenges, adaptation must be taken into account. Astronaut participation in extended-duration missions, the prospects of a return to the Moon, continuing public enthusiasm for a mission to Mars, the reformulation of research questions following the publication of *Safe Passage*, and the coevolution of NASA's Bioastronautics Critical Path Roadmap and the National Space Biomedical Research Institute initiated a new era for psychology. According to our analysis, since the dawn of the modern space station era, there has been an increase in both research and operational interest in spaceflight behavioral health. Slowly, and perhaps painfully, psychology has gained greater recognition within the U.S. space program, and there is a growing convergence of interests to target research at operational problems.[82]

Current NASA administration has mandated that human research be operationally relevant. This is partly driven by funding shortages and partly by needs to meet NASA performance standards and requirements when astronauts once again venture beyond low-Earth orbit. The new Human Research Program documents including the "Human Research Program Requirements Document" and the "Human Research Program Integrated Research Plan" are the bases for defining, documenting, and allocating human research program requirements as they have evolved from the older Bioastronautics Critical Path Roadmap and new NASA standards and requirements that emphasize future missions. As explained on the NASA Web site, "The Human Research Program (HRP) delivers human health and performance countermeasures, knowledge, technologies, and tools to enable safe, reliable, and productive human space exploration. This Integrated Research Plan (IRP) describes the program's research activities that are intended to address the needs of human space exploration and serve IRP customers. The timescale of human space exploration is envisioned to take many decades. The IRP illus-

---

82. Albert A. Harrison, "Behavioral Health: Integrating Research and Application in Support of Exploration Missions," *Aviation, Space, and Environmental Medicine* 76, no. 6, sect. II (June 2005): B3–B12.

trates the program's research plan through the timescale of early lunar missions of extended duration."[83]

We can see the preliminary outlines of a comprehensive and continuing program in spaceflight behavioral health. A comprehensive program in spaceflight behavioral health will have to be broad-based; be interdisciplinary; and address issues at the individual, small-group, and organizational levels. It will require multiple, convergent methods including archival research, field observations, and both field and laboratory experiments. Research falling under this umbrella must meet high scientific standards, achieve flight certification, and be palatable to astronauts. Only with continued interest and support from NASA—and from psychologists—will spaceflight behavioral health flourish. Long-term success will require accessible, peer-reviewed publications and efforts to target young investigators to replace those who retire. An ongoing behavioral database could prove very useful. For over 15 years, David Musson, Robert Helmreich, and their associates have been developing a database that includes astronauts as well as professionals who work in other demanding environments.[84] As they point out, this kind of database provides many opportunities for studies in such areas as the effectiveness of recruiting and selection procedures, performance changes over time, and attrition.

Psychology is in a better position to be of help. Many of the theories and tools that are proving useful today were not available at the dawn of the Space Age. New (relative to 1960) resources include cognitive models, which emphasize our information processing power, and humanistic or "positive psychology" models that stress people's positive, striving nature.[85] These new models have allowed psychologists a fresh take on many important issues. Human factors psychologists benefit from modern computer modeling technologies and increasing evidence of the importance of taking the person into account when developing a human or human-robotic system.

---

83. NASA Johnson Space Center, *Human Research Program Integrated Research Plan*, Supplement A1, *Behavioral Health and Performance*, 2008, available at *http://humanresearch.jsc.nasa.gov/elements/smo/docs/bhp_irp_supplemental_v1.pdf* (accessed 21 May 2010).

84. D. M. Musson and R. L. Helmreich, " Long-Term Personality Data Collection in Support of Spaceflight Analogue Research," *Aviation, Space, and Environmental Medicine* 76, no. 6, sect. II (2005): B119–B125.

85. Suedfeld, "Invulnerability, Coping, Salutogenesis, Integration."

Research technology has changed dramatically over the past 50 years, and the new technology has also been useful for increasing psychology's contributions to NASA. These changes are evident wherever we look, from questionnaire construction to data analysis. Today, miniaturization and computer technology enable psychological assessments and evidence-based countermeasures that would have been impossible in the 1960s.

Minimally intrusive techniques are particularly useful, and one of these is based on nonintrusive computer monitoring of facial expression.[86] Another approach is monitoring cognitive functioning through computer analysis of speech.[87] Encouraging astronauts to monitor their own behavior reduces the threat that performance lapses could lead to flight disqualification. This self-monitoring has been accomplished by means of computers and personal digital assistants (PDAs) that are programmed to measure several dimensions of cognitive functioning (attention, information processing, and recall). Astronauts may use the results of these tests to gauge their own preparedness to engage in a particular activity.[88]

While we see evidence of an expanding role, our profession's future in spaceflight is by no means assured. NASA's resistance to psychology is by no means fully overcome. NASA Administrators must still concern themselves with public relations. Project managers and engineers must still get on with their tasks within the real constraints of cost and practicality. Astronauts remain sensitive to possible threats to flight assignments and careers. The focusing events of *Mir* and the ISS were less than two decades ago, and it is too early to tell if the new interest and infrastructure can withstand the vagaries of funding variations or national and organizational politics.

---

86. D. F. Dinges, R. L. Rider, J. Dorrian, E. L. McGlinchey, N. L. Rogers, Z. Cizman, S. K. Goldenstein, C. Vogler, S. Venkartamarian, and D. N. Metaxas, "Optical Computer Recognition of Facial Expressions Associated with Stress Induced by Performance Demands," *Aviation, Space, and Environmental Medicine* 76, no. 6, sect. II (June 2005): B172–182.

87. P. Lieberman, A. Morey, J. Hochstadt, M. Larson, and S. Mather, "Mount Everest: A Space Analogue for Speech Monitoring of Cognitive Deficits and Stress," *Aviation, Space, and Environmental Medicine* 76, no. 6, sect. II (June 2005): B198–B207.

88. J. M. Shephard and S. M. Kosslyn, "The MiniCog Rapid Assessment Battery: A 'Blood Pressure Cuff' for the Mind," *Aviation, Space, and Environmental Medicine* 76, no. 6, sect. II (June 2005): B192–B197.

Chapter 3

From Earth Analogs to Space: Getting There from Here

Sheryl L. Bishop

    Department of Preventive Medicine and Community Health and School of Nursing
    University of Texas Medical Branch

ABSTRACT

The need to find relevant terrestrial substitutes, that is, analogs, for teams operating in extraterrestrial and microgravity environments is driven by extraordinary demands for mission success. Unlike past frontiers where failure on the part of various groups to succeed represented far more limited implications for continued progress within these environments, accidents like *Challenger* in 1986 and *Columbia* in 2003 underscored the magnified cost of failure for space missions. Where past human frontiers were characterized by centralized decisions to engage in exploration and development largely under the dictates of authoritarian governments or individual sponsors, the exploration of space has been significantly influenced by the general public's perception of "acceptable risk" and fiscal worthiness. To date, space missions have failed due to technological deficiencies. However, history is replete with examples of exploration and colonization that failed due to human frailties, including those that reflect failures of the group. Both historical literature and research on teams operating within extreme environments, including space, have clearly indicated that psychological and sociocultural factors are components critical for individual and group success. Given the limited access to the space frontier and the investment in collective effort and resources, our ability to study individual and group functioning in the actual space environment has been, and will continue to be, severely limited. Thus, studying groups in terrestrial extreme environments as analogs has been sought to provide *predictive* insight into the many factors that impact group performance, health, and well-being in challenging environments.

    This chapter provides an overview of the evolution of research utilizing terrestrial analogs and addresses the challenges for selecting, training, and supporting teams for long-duration space missions. An examination of how analog environments can contribute to our knowledge of factors affecting functioning and well-being at both the physiological and the psychological levels will help define the focus for future research.

INTRODUCTION

Humans have long speculated about, studied, and striven to explore the heavens. Many of our earliest myths, such as the flight of Daedalus and Icarus too close to the Sun on wings made of wax, expressed our desire to explore beyond the boundaries of Earth as well as our willingness to push current technology to its limits. Considerations by the earliest philosophers and scientists, including Archimedes, Galileo Galilei, Nicolaus Copernicus, Leonardo da Vinci, Sir Isaac Newton, Jules Verne, H. G. Wells, or Percival Lowell, eventually generated a whole new genre of fictional literature built upon scientific extrapolations, dubbed "science fiction," and gave voice to their speculations about the nature of extraterrestrial environments. Modern scientists and pioneers led by the Wright brothers, Robert Goddard, Konstantin Tsiolkovsky, Hermann Oberth, Wernher von Braun, Sergey Korolev, Yuri Gagarin, and Neil Armstrong pushed the boundaries of knowledge about flight and extended human inquiry beyond our terrestrial boundaries into our local and extended galactic neighborhood. For serious considerations of how humans will fare in space, we have had to extrapolate from human experience on Earth in environments that challenge us in, ideally, similar ways. However, the search for space analog environments in which to systematically study individual and group adaptation has had to grapple with some significant limitations, i.e., the impossibility of a substitute for a microgravity or reduced-gravity environment or environments that holistically mimic radiation profiles and their inherent danger for those beyond Earth's magnetic field. Since there is no direct equivalent for space, all analog environments are simulations of greater or lesser fidelity along varying dimensions of interest. Some analog environments provide extremely good characterizations of expected challenges in testing equipment or hardware, e.g., environmental chambers such as the Space Shuttle mock-ups of the various decks or the cargo bay in NASA's Weightless Environmental Training Facility (WET-F), but lack any relevance to assessing how human operators will fare psychologically or as a team. Others, like chamber studies, address important components of human adaptation, e.g., confinement, but fail utterly to incorporate true environmental threats. Others allow for the impact of true dangerous, unpredictable environments but lack any way to systematically compare across specific environments. The spectrum of fidelity to space among terrestrial analogs ranges from laboratory studies where the impact of environmental threat and physical hardship, as well as true isolation and confinement, are limited and, even, sometimes absent, to real teams in real, extreme environments characterized by very little control over extraneous variables.

This, then, is the challenge. Unlike the testing of hardware, where various components can be reliably evaluated separately, the study of humans, and teams in particular, is a dynamic endeavor requiring in situ study of the collective. To develop reliable protocols based on empirical evidence to select, monitor, and support teams effectively in space necessarily involves the demand to study teams in analog environments that replicate a wide range of physiological, psychological, and psychosocial factors interacting both with the environment and within the team. The high degree of reliance on technology for life support, task performance, and communication must be integrated with new measurement methodologies to overcome heretofore intrusive measurement modalities. The growing frequency of multinational and multicultural teams and the demand for longer-duration missions both further compound the complexity of the challenge. While the primary goal has been the insurance of human health and well-being, the expectation has been that such priorities will naturally lead to improved chances for performance and mission success. Yet achieving this goal depends largely on how well our analogs prepare us for living and working in space.

Analogs for human individual and group performance in space has involved two basic approaches: 1) constructing an environment within a laboratory setting with maximum control over extraneous variables and utilizing volunteer research subjects or 2) studying naturally occurring real-world groups in real environments characterized by a number of confounds.[1] Each comes with its own limitations and strengths. In any evaluation of the value of the analog, the pros and cons of each environment need to

---

1. W. Haythorn and I. Altman, "Personality Factors in Isolated Environments," in *Psychological Stress: Issues in Research*, ed. M. Trumbull (New York: Appleton-Century-Crofts, 1966); J. P. Zubek, *Sensory Deprivation: Fifteen Years of Research* (New York: Appleton-Century-Crofts, 1969).

be kept in mind. This is especially true when assessing the generalizability of insight of psychosocial factors from substitute environments for space.

Before we began deliberately constructing controlled laboratory environments, there were the records of early expeditionary explorations into various places on Earth.[2] The tradition of publishing personal diaries and mission recounts has been similarly observed by the earliest explorers of space.[3] Secondary analyses of historical expeditions have become increasingly popular in recent years.[4] The very character of natural environments typically guarantees that there will be at least some, if not substantial, periods of inaccessibility, lack of communication or contact, little accessibility of real-time support, and great demands on individuals and groups to engage in autonomous decision-making, problem-solving, conflict resolution, self-monitoring, and self-regulation. These demands inherently build in the potential for conflict with external mission support personnel and researchers who find adherence to mission schedules and timelines far easier to maintain than do those actually on the mission. Shared perspective between these groups becomes increasingly difficult to promote as mission duration, distance, and environmental demands play larger roles in daily decisions of the teams than do seemingly arbitrary mission schedules.

Measurement of these factors is compromised as teams become preoccupied with dealing with the environment, become antagonistic to external evaluation, become noncompliant with schedules that become unimportant to participants, and engage in a general reprioritization of activities that emphasizes near-term, more salient goals (e.g., personal comfort, leisure) over and above long-term mission goals (e.g., study data). Such difficulties have raised questions about the worth of studying groups in real-world environments. In actuality, these conditions are *exactly* what is needed to simulate space missions that have grown in duration,

---

2. A. Greely, *Three Years of Arctic Service: An Account of the Lady Franklin Bay Expedition of 1881–1884, and the Attainment of the Farthest North* (New York: Scribner, 1886); V. Stefansson, *The Adventure of Wrangel Island* (New York: MacMillan Company, 1925); R. Pearce, "Marooned in the Arctic: Diary of the Dominion Explorers' Expedition to the Arctic, August to December 1929," *Northern Miner* (Winnipeg, MB, 1930).

3. V. Lebedev, *Diary of a Cosmonaut: 211 Days in Space* (College Station, TX: Phytoresource Research, Inc., 1988); J. Lovell and J. Kluger, *Apollo 13 [Lost Moon: The Perilous Voyage of Apollo 13]* (New York: Pocket Books, 1994).

4. J. Stuster, *Bold Endeavors* (Annapolis, MD: Naval Institute Press, 1996).

distance from Earth, complexity, and challenge. However, space missions will also be, at least for the foreseeable future, characterized by an extraordinary degree of control, from selecting who goes to establishing the daily details of mission tasks and schedules—elements that are far more variable in real-world groups, such as those in Antarctica or part of polar or mountaineering expeditions. In real-world groups that have higher degrees of structure and control, such as military teams, the command and control structure is distinctly different from the current scientist-astronaut organizational structure of space missions. Fundamental differences in group structures, such as leadership and authority, represent significant elements in whether findings from terrestrial analogs translate to future space crews.

The need for control over the inherent chaos of real-world environments in order to definitively identify critical factors that affect individual and group performance was the driver behind the development of constructed environments of various complexities. Useful data from such artificial environments depend on whether participants are truly immersed in the fiction of a simulation and are responding in the same way they would if the environment were real. This is the paradox researchers in analog environments face: In laboratory studies, the very attributes of the environment that have the greatest impact on performance are removed (e.g., real danger, uncontrolled events, situational ambiguity, uncertainty, or the interaction with the extreme environment itself). If these features are compromised, as many have argued, then is there value in conducting such laboratory studies?[5] On the pro side, laboratory chamber studies have provided opportunities to evaluate methods of monitoring psychological and interpersonal parameters for subsequent application during real flights and have identified issues that might cause psychological and interpersonal problems in space. They have also provided empirical evidence for a number of behavioral issues anecdotally reported from space, e.g., the tendency of crews to direct aggression toward personnel at Mission Control.[6] They

---

5. L. A. Palinkas, "On the ICE: Individual and Group Adaptation in Antarctica," 2003, available at *http://www.sscnet.ucla.edu/anthro/bec/papers/Palinkas_On_The_Ice.pdf* (accessed 12 June 2007); P. Suedfeld, "What Can Abnormal Environments Tell Us About Normal People? Polar Stations as Natural Psychological Laboratories," *Journal of Environmental Psychology* 18 (1998): 95.

6. N. Kanas, V. Salnitskiy, E. M. Grund, et al., "Social and Cultural Issues During Shuttle/Mir Space Missions," *Acta Astronautica* 47 (2000): 647; G. M. Sandal, R. Vaernes, and H. Ursin, "Interpersonal Relations During Simulated Space Missions," *Aviation, Space, and*

are well suited to first-line inquiry when there is a need to investigate the characteristics of a particular phenomenon suspected of being present. However, complexity is a key defining trait of stressed operational environments. Total reliance on laboratory studies and the presumption of broad generalizability, particularly for research on high-stress, high-risk environments, is highly likely to lead to dissociation between actual operational findings and laboratory and experimental studies.[7] Conversely, data on real-world groups situated in extreme environments has provided insight into a host of factors that impact group performance, health, and well-being emergent from the interaction between the individual, the team, and the environment. The differences found between studies conducted in experimentally controlled chambers and those conducted in messy, noisy, in situ real environments appears to be due to the critical presence of real environmental threat and physical hardship, as well as true isolation and confinement, which have proven to be key factors in individual and group coping. Additionally, when comparing extreme environments with non-extreme natural environments in which people normally operate, the level, intensity, rate of change, and diversity of physical and social stimuli, as well as behavior settings and possible behaviors within an extreme environment, are far more restricted.[8]

Thus, real teams in extreme environments have validated or corrected findings from chamber studies where critical environmental factors are typically absent or blunted. Real extreme environments allow us to examine various aspects of the psychophysiological relationship that are essential to fully understanding the adaptation

---

Environmental Medicine 66 (1995): 617; V. I. Gushin, V. A. Kolintchenko, V. A. Efimov, and C. Davies, "Psychological Evaluation and Support During EXEMSI," in *Advances in Space Biology and Medicine*, ed. S. Bonting (London: JAI Press, Inc., 1996), p. 283; V. I. Gushin, T. B. Zaprisa, V. A. Kolintchenko, A. Efimov, T. M. Smirnova, A. G. Vinokhodova, and N. Kanas, "Content Analysis of the Crew Communication with External Communicants Under Prolonged Isolation," *Aviation, Space, and Environmental Medicine* 12 (1997): 1093.

7. A. D. Baddeley, "Selecting Attention and Performance in Dangerous Environments," *British Journal of Psychology* 63 (1972): 537; G. W. McCarthy, "Operational Relevance of Aeromedical Laboratory Research," abstract no. 24 (paper presented as part of the Aerospace Medical Association's 69th Annual Scientific Meeting, Seattle, WA, 17–21 May 1988), p. 57; J. D. Mears and P. J. Cleary, "Anxiety as a Factor in Underwater Performance," *Ergonomics* 23, no. 6 (1980): 549; G. Wilson, J. Skelly, and B. Purvis, "Reactions to Emergency Situations in Actual and Simulated Flight" (presented as a paper at the Aerospace Medical Panel Symposium, The Hague, Netherlands, 1989).

8. Suedfeld, "What Can Abnormal Environments Tell Us About Normal People?": 95.

of humans to the stresses of these environments and, ultimately, to space. Space, of course, will be the final testing ground for our accumulated knowledge. But are we stuck with choosing between chamber studies and naturally occurring opportunistic teams in real extreme environments? A more recent, hybrid approach of situating research facilities within extreme environments offers a good compromise between the artificial conditions of the laboratory and the open-ended, full access of an expeditionary mission. When teams or individuals operate in extreme environments, their responses are more purely a product of either situational drivers or internal personal characteristics. To the extent that an extreme environment is well characterized and known, it gains in fidelity and allows more accurate inferences about key phenomena to be drawn. For these very reasons, Palinkas has strongly argued that the cumulative experience with year-round presence in Antarctica makes it an ideal laboratory for investigating the impact of seasonal variation on behavior, gaining understanding about how biological mechanisms and psychological processes interact, and allowing us to look at a variety of health and adaptation effects.[9]

PSYCHOLOGY AND SPACE

One important fact, which has emerged during decades of research, is that in the study of capsule environments there are few main effect variables. Almost every outcome is due to an interaction among a host of physical and social environmental variables and personality factors. Thus, although we conceptually deconstruct the situation into particular sources of variance, we must remember that how people experience an environment is more important than the objective characteristics of the environment.[10]

Investigations into psychological and psychosocial adaptation to extreme environments as substitutes for space are recent phenomena. Expeditions and forays

---

9. Palinkas, "On the ICE."
10. P. Suedfeld and G. D. Steel, "The Environmental Psychology of Capsule Habitats," *Annual Review of Psychology* 51 (2000): 230.

into these environments have historically been for the purposes of exploration, and the primary metric of successful adaptation was survival. One could argue that chronicles such as the *Iliad* and the *Odyssey* were early examples of more recent diaries such as those that recounted the historic race to reach the South Pole between modern polar expeditions lead by Roald Amundsen, who reached the South Pole in 1911, and Robert F. Scott, who reached the South Pole in 1912. Humans have been periodically living and working in Antarctica, one of the most challenging environments on Earth, for over a hundred years. The first winter-over in Antarctica occurred during 1898–99 on board an icebound ship, the *Belgica*, on which Amundsen served as a second mate. A continuous presence on our furthermost southern continent has only been in place since the International Geophysical Year of 1956–57. Systematic research on isolated, confined environments can arguably be dated as beginning as recently as the late 1950s by the military, and much of the early work focused on purely physiological parameters. In their seminal collection of papers dealing with isolated environments from Antarctica to outer space, A. A. Harrison et al. pointed out that early work on psychological factors in extreme environments is often recounted as beginning with C. S. Mullin's research on states of consciousness; E. K. E. Gunderson and colleagues' comprehensive work on adaptation to Antarctica; and classic laboratory studies on group dynamics conducted by I. Altman, W. W. Haythorn, and associates.[11]

Regardless of which analog is used to understand what helps or hinders individuals and groups in functioning well under extreme environmental challenges, it is necessary to characterize what we need to know for space. Although specific conditions of the setting vary, most extreme environments share common characteristics: 1) a high reliance on technology for life support and task performance; 2) notable degrees of physical and social isolation and confinement; 3) inherent high risks

---

11. A. A. Harrison, Y. A. Clearwater, and C. P. McKay, *From Antarctica to Outer Space: Life in Isolation and Confinement* (New York: Springer-Verlag, 1991); C. S. Mullin, "Some Psychological Aspects of Isolated Antarctic Living," *American Journal of Psychiatry* 111 (1960): 323; E. K. E. Gunderson, "Individual Behavior in Confined or Isolated Groups," in *Man in Isolation and Confinement*, ed. J. Rasmussen (Chicago: Aldine, 1973), p. 145; E. K. E. Gunderson, "Psychological Studies in Antarctica," in *Human Adaptability to Antarctic Conditions*, ed. E. K. E. Gunderson (Washington, DC: American Geophysical Union, 1974), p. 115; I. Altman, "An Ecological Approach to the Functioning of Isolated and Confined Groups," in *Man in Isolation and Confinement*, ed. Rasmussen, p. 241; W. W. Haythorn, "The Miniworld of Isolation: Laboratory Studies," in *Man in Isolation and Confinement*, ed. Rasmussen, p. 219.

and associated costs of failure; 4) high physical/physiological, psychological, psychosocial, and cognitive demands; 5) multiple critical interfaces (human-human, human-technology, and human-environment); and 6) critical requirements for team coordination, cooperation, and communication.[12] This last is not insignificant. The accumulated knowledge to date is still fairly rudimentary, given the short historical emergence of the "Space Age." Drawing on research from a number of fields (e.g., social psychology, human factors, military science, management, anthropology, and sociology), researchers easily identified a number of factors that need further investigation. As early as the 1980s, psychological and sociocultural issues had been acknowledged by the National Commission on Space (1986), the National Science Board (1987), and the Space Science Board (1987) to be critical components to mission success, as robust evidence from Antarctica clearly showed psychological issues to impact human behavior and performance significantly in most challenging environments, especially those characterized by isolation and confinement.[13] Studies in a variety of analog environments, e.g., Antarctica, underwater capsules, submarines, caving and polar expeditions, and chamber studies, have confirmed that mission parameters have a significant influence upon the type of "best-fit" crew needed and have isolated a number of psychosocial issues that may negatively affect crewmembers during multinational space missions.[14] These issues include 1) tension resulting

---

12. S. L. Bishop, "Psychological and Psychosocial Health and Well-Being at Pole Station," in *Project Boreas: A Station for the Martian Geographic North Pole*, ed. Charles S. Cockell (London: British Interplanetary Society, 2006), p. 160.

13. National Science Board, *The Role of the National Science Foundation in Polar Regions* (Washington, DC: National Academy of Sciences, 1987); Space Science Board, *A Strategy for Space Biology and Medical Science* (Washington, DC: National Academy Press, 1987); National Commission on Space, *Pioneering the Space Frontier* (New York: Bantam Books, 1986).

14. L. A. Palinkas, E. K. E. Gunderson, and R. Burr, "Social, Psychological, and Environmental Influences on Health and Well-Being of Antarctic Winter-Over Personnel," *Antarctic Journal of the United States* 24 (1989): 207; L. A. Palinkas, "Sociocultural Influences on Psychosocial Adjustment in Antarctica," *Medical Anthropology* 10 (1989): 235; L. A. Palinkas, "Psychosocial Effects of Adjustment in Antarctica: Lessons for Long-Duration Spaceflight," *Journal of Spacecraft* 27, no. 5 (1990): 471; L. A. Palinkas, "Effects of Physical and Social Environments on the Health and Well Being of Antarctic Winter-Over Personnel," *Environment* 23 (1991): 782; C. Anderson, "Polar Psychology—Coping With It All," *Nature* 350, no. 6316 (28 March 1991): 290; H. Ursin, "Psychobiological Studies of Individuals in Small Isolated Groups in the Antarctic and Space Analogue," *Environment and Behavior* 6 (23 November 1991): 766; L. Palinkas, E. K. E. Gunderson, and A. W. Holland, "Predictors of Behavior and Performance in Extreme Environments: The Antarctic Space Analogue Program," *Aviation, Space, and*

from external stress, 2) factors related to crew heterogeneity (e.g., differences in personality, gender, and career motivation); 3) variability in the cohesion of the crew; 4) improper use of leadership role (e.g., task/instrumental versus emotional/ supportive); 5) cultural differences; and 6) language differences. Of particular uniqueness to challenging environments is the fact that successful performance requires *competent* team interaction, including coordination, communication, and cooperation. The functioning of the operational team often determines the success or failure of the mission. Experience in spaceflight, aviation, polar, and other domains indicates that the stressors present in extreme environments, such as fatigue, physical danger, interpersonal conflict, automation complexity, risk, and confusion, often challenge team processes. The contribution of interpersonal and intrapersonal factors is substantial. For instance, a robust body of evidence from both civilian and military aviation identifies the majority of aircraft accidents as due to human and crew-related performance factors.[15] Analyses of critical incidents in medical operating

---

Environmental Medicine 71 (2000): 619; S. L. Bishop and L. Primeau, "Assessment of Group Dynamics, Psychological and Physiological Parameters During Polar Winter-Over" (paper presented as part of the Human Systems Conference, Nassau Bay, TX, 20–22 June 2001); L. Palinkas, "The Psychology of Isolated and Confined Environments: Understanding Human Behavior in Antarctica," *American Psychologist* 58, no. 5 (2003): 353; R. H. Gilluly, "Tektite: Unique Observations of Men Under Stress," *Science News* 94 (1970): 400; J. L. Sexner, "An Experience in Submarine Psychiatry," *American Journal of Psychiatry* 1 (1968): 25; G. M. Sandal, I. M. Endresen, R. Vaernes, and H. Ursin, "Personality and Coping Strategies During Submarine Missions," *Military Psychology* 11 (1999): 381; S. L. Bishop, P. A. Santy, and D. Faulk, "Team Dynamics Analysis of the Huautla Cave Diving Expedition: A Case Study," *Human Performance and Extreme Environments* 1, no. 3 (September 1998): 34; G. M. Sandal, R. Vaernes, P. T. Bergan, M. Warncke, and H. Ursin, "Psychological Reactions During Polar Expeditions and Isolation in Hyperbaric Chambers," *Aviation, Space, and Environmental Medicine* 67, no. 3 (1996): 227; S. L. Bishop, L. C. Grobler, and O. Schjøll, "Relationship of Psychological and Physiological Parameters During an Arctic Ski Expedition," *Acta Astronautica* 49 (2001): 261; N. Kanas, "Psychosocial Factors Affecting Simulated and Actual Space Missions," *Aviation, Space, and Environmental Medicine* 56 (1985): 806.

15. The Boeing Company, "Statistical Summary of Commercial Jet Aircraft Accidents: Worldwide Operations, 1959–1993," in *Boeing Airplane Safety Engineering Report B-210B* (Seattle, WA: Boeing Commercial Airplane Group, 1994); M. W. Raymond and R. Moser, "Aviators at Risk," *Aviation, Space, and Environmental Medicine* 66, no. 1 (1995): 35; D. S. Ricketson, W. R. Brown, and K. N. Graham, "3W Approach to the Investigation, Analysis, and Prevention of Human-Error Aircraft Accidents," *Aviation, Space, and Environmental Medicine* 51 (1980): 1036; B. L. Weiner, B. O. Kanki, and R. L. Helmreich, *Cockpit Resource Management* (New York: Academic Press, 1993); D. A. Wiegmann and S. A. Shappel, "Human Factors Analysis of Postaccident Data: Applying Theoretical Taxonomies of Human

rooms indicate that 70 to 80 percent of medical mishaps are due to team and interpersonal interactions among the operating room team.[16] From pilot to surgeon, firefighter, polar expeditioner or astronaut, we need to know if the characteristics that define adaptable and functional individuals and teams have commonalities across various environments. It is therefore critical that teamwork in these environments be examined and understood. A fundamental need to enable these investigations is developing reliable, minimally intrusive and valid methodologies for assessing individual and group responses to these stressors and identifying dysfunctional and functional coping responses. The use of extreme environments with characteristics relevant to those inherent in space travel and habitation will play a crucial role in preparing humans for egress from planet Earth.

Given the disparate nature of these various environments, Peter Suedfeld has proposed five key principles that may be useful guides in assessing the relevance of various extreme environments as viable analogs for space or providing the basis for cross-comparisons:

Principle 1: Researchers should think in terms of experiences within environments rather than of environmental characteristics;

Principle 2: Researchers should study differences and similarities between experiences, which are not the same as those between environments;

Principle 3: Analogies should be based on similarities of experience, not necessarily of environment;

Principle 4: Research should look at systematic links between personality factors and experience; and

Principle 5: Experience is continuous and integrated.[17]

Error," *International Journal of Aviation Psychology* 7 (1997): 67; D. W. Yacovone, "Mishap Trends and Cause Factors in Naval Aviation: A Review of Naval Safety Center Data, 1986–1990," *Aviation, Space, and Environmental Medicine* 64 (1993): 392.

16. B. Sexton, S. Marsch, R. Helmreich, D. Betzendoerfer, T. Kocher, and D. Scheidegger, "Jumpseating in the Operating Room," *Journal of Human Performance in Extreme Environments* 1, no. 2 (1996): 36; J. A. Williamson, R. K. Webb, A. Sellen, W. B. Runciman, and J. H. van der Walt, "Human Failure: An Analysis of 2000 Incident Report," *Anesthesia Intensive Care* 21 (1993): 678.

17. P. Suedfeld, "Groups in Isolation and Confinement: Environments and Experiences," in *From Antarctica to Outer Space: Life in Isolation and Confinement*, ed. A. A. Harrison, Y. A. Clearwater, and C. P. McKay (New York: Springer-Verlag, 1991), p. 135.

CRITICAL PSYCHOSOCIAL ISSUES FOR SPACE

The research on teams has, to date, focused on and identified needs for further research under four broad categories. The intent here is not to recite the spectrum of findings across analogs within these areas, but to articulate how analog environments can address these areas.

- *Selection issues* deal with the evaluation of existing ability, trainability, and adaptability of prospective team members. It is not merely a matter of selecting-out pathological tendencies, but, as importantly, selecting-in desirable characteristics. How can analog environments allow us to investigate the impact of various individual and group characteristics upon individual and group performance?
- The impact of *isolation and confinement* has been shown to be significantly impacted by various moderator variables, e.g., the difficulty of rescue. While an emergency on the International Space Station certainly poses difficulties regarding time to rescue, one can argue that the difficulties inherent in a Mars mission or even here on Earth from the Antarctic in midwinter, where weather conditions may absolutely make rescue impossible for long periods, carry a qualitatively different psychological impact. An emergency on a mission to Mars will preclude any chance of rescue and necessitate a high degree of autonomy for the crew in making decisions without any real-time mission support. The degree to which such factors magnify the negative effects of isolation and confinement is critical to assess.
- *Group interaction and group processes* are not a simple sum of the individuals that make up the group. Complex interactions can reinforce, undermine, or create new behaviors in the individuals involved. Identification of group fusion (factors that encourage group cohesion) and fission (factors that contribute to group conflict) variables are elementary to creating habitats and work schedules, composing groups, and a myriad of other factors that will enable groups to function effectively and ensure individual and group well-being. For instance, in a study of Antarctic winter-over personnel, Palinkas found that personnel at Palmer (a small station) spent 60 percent of their waking hours alone and retreated to their bedrooms extensively for privacy. These behaviors could be considered fission factors as they promote withdrawal, social isolation, and distancing from one's teammates. On the other hand, if the use of privacy served to control the amount of contact and decreased tensions and group conflict, they would be

considered fusion factors. He also found that intermittent communication was a major source of conflict and misunderstanding between crews and external support personnel, a clear source of fission influence. Examples of fusion factors for this group were effective leadership styles, which played a significant role in station and crew functioning, as well as the ability to move furniture and decorate both common and private areas, which facilitated adaptation and adjustment.[18]

- *Individual and crew performance* is perhaps the clearest, most frequently studied outcome. Yet there are challenges in defining what constitutes acceptable outcomes at both the individual and group levels. They are not always the same thing, as investigations into missions that failed to meet expectations have repeatedly confirmed. It is a mistake to try to assess and maximize performance without understanding group dynamics, the effects of isolation and confinement or the environment in general on inhabitants. Given that our selection criteria have been little more than ruling out pathology and matching task requirements with technical proficiency within individuals, it is of little surprise that our efforts to implement performance improvements have been only modestly successful and fraught with inconsistent results. It is necessary to take the next steps to identify which individual and group characteristics are maximally associated with adaptation and functioning in these high-challenge environments.

TERRESTRIAL ANALOGS FOR SPACE

There are surprising similarities and differences found across environments. G. M. Sandal et al. found that coping strategies during confinement on polar expeditions were different from those in hyperbaric chambers.[19] Whereas polar teams evidenced a delay interval with a marked drop in aggression until after the first quarter, with concomitant increase in homesickness, chamber teams displayed a steady gradual increase in coping over time. A number of researchers have noted that it is not the site that seems to matter, but rather it is the differences in the mis-

---

18. Palinkas, "Psychosocial Effects of Adjustment in Antarctica: Lessons for Long-Duration Spaceflight": 471.
19. Sandal, Vaernes, Bergan, Warncke, and Ursin, "Psychological Reactions During Polar Expeditions and Isolation in Hyperbaric Chambers": 227.

sion profiles, e.g., tasks (daily achievement of a distance goal versus stationkeeping) or duration (short versus long).

In fact, studies addressing Suedfeld's Principle 4 investigating personality characteristics have produced supporting evidence for a focus on the experience as the defining factor rather than the environment per se. The most persistently investigated personality assessment for the last 15 years has been the NEO-PI by P. T. Costa and R. R. McCrae.[20] This instrument assesses five global dimensions of personality: neuroticism, extraversion, openness to experience, agreeableness, and conscientiousness. These dimensions have been found to be associated with the previous personality "right stuff/wrong stuff/no stuff" profiles identified by Helmreich et al. in longitudinal studies of American astronaut candidate performance.[21] Additionally, measures of achievement motivation, interpersonal orientation, Type A, stress, and coping have been frequently evaluated. Recent studies have found evidence that agreeableness and conscientiousness seem to better predict performance at the global level, along with specific facets of extraversion.[22] Conscientiousness, extraversion, and agreeableness have been found to be related more strongly to constructive change-oriented communication and cooperative behavior than to task performance. Cognitive ability appears to be related more strongly to task performance than to constructive change-oriented communication or cooperative behavior. Results also demonstrate contrasting relationships for agreeableness (positive with cooperative behavior and negative with constructive change-oriented communication).[23] However, another personal-

---

20. P. T. Costa, Jr., and R. R. McCrae, *NEO Five-Factor Inventory* (Lutz, FL: Psychological Assessment Resources, Inc., 1978, 1985, 1989, 1991).

21. T. J. McFadden, R. Helmreich, R. M. Rose, and L. F. Fogg, "Predicting Astronaut Effectiveness: A Multivariate Approach," *Aviation, Space, and Environmental Medicine* 65 (1994): 904.

22. P. Suedfeld and G. D. Steel, "The Environmental Psychology of Capsule Habitats," *Annual Review of Psychology* 51 (2000): 227; R. M. Rose, R. L. Helmreich, L. F. Fogg, and T. McFadden, "Psychological Predictors of Astronaut Effectiveness," *Aviation, Space, and Environmental Medicine* 64 (1994): 910; R. R. McCrae and J. Allik, *The Five-Factor Model of Personality Across Cultures* (Dordrecht, Netherlands: Kluwer, 2002).

23. M. R. Barrick, G. L. Stewart, M. J. Neubert, and M. K. Mount, "Relating Member Ability and Personality to Work-Team Processes and Team Effectiveness," *Journal of Applied Psychology* 83 (1998): 377; L. Ferguson, D. James, F. O'Hehir, and A. Sanders, "Pilot Study of the Roles of Personality, References, and Personal Statements in Relation to Performance over the Five Years of a Medical Degree," *British Medical Journal* 326, no. 7386 (22 February 2003): 429; J. A. LePine, "Team Adaptation and Postchange Performance: Effects of

ity cluster has been identified in studies of successful polar trekking groups that is distinctly different from the "right stuff" profile in which factors indicative of individuals who are loners seem to be supportive of adaptation, i.e., happier alone than dependent on others, highly autonomous, independent, uncomfortable about and relatively uninterested in accommodating others in a group, task-oriented and somewhat competitive.[24] Since we do not have enough data to reliably draw inferences about these individuals, it is mere speculation at this time that perhaps the intense task focus of a polar trek, in which each individual is highly autonomous and individually self-reliant during the long travel each day, situated in an environment that precludes group interaction except for fundamental coordination of locomotion across the terrain, *selects for* individuals that are distinctly different from those who would occupy a habitat or confined environment for long durations. In other words, only individuals with this inward, self-focused personality would find such challenges rewarding and be successful at these tasks. Similarly, an apparently adaptive personality profile has emerged from winter-overers that is characterized by low levels of neuroticism, desire for affection, boredom, and need for order, as well as a high tolerance for lack of achievement, which would fit well in an environment where isolation and confinement prevented accomplishments and the participants experienced frequent shortages and problems.[25] Those that would best adapt would be those who could more quickly adjust their expectations to the immediate situation and tolerate such obstacles. If this hypothesis is substantiated, then we must carefully match the characteristics of the individual to the environment as well as the group in order to maximize successful adaptation and performance.

Psychological research to date seems to support two general findings: 1) there do seem to be consistencies in the personality profile of functional and dysfunctional teams, and 2) characteristics of the mission may define very different personality

---

Team Composition in Terms of Members' Cognitive Ability and Personality," *Journal of Applied Psychology* 88, no. 1 (February 2003): 27; T. A. Judge and R. Ilies, "Relationship of Personality to Performance Motivation: A Meta-Analytic Review," *Journal of Applied Psychology* 87, no. 4 (August 2002): 797.

24. E. Rosnet, C. Le Scanff, and M. Sagal, "How Self-Image and Personality Affect Performance in an Isolated Environment," *Environmental Behavior* 32 (2000): 18.

25. L. Palinkas, E. K. E. Gunderson, and A. W. Holland, "Predictors of Behavior and Performance in Extreme Environments: The Antarctic Space Analogue Program," *Aviation, Space, and Environmental Medicine* 71 (2000): 619.

profiles as best fit. Insomuch as it is possible to select for hardier and better-fit personalities by filtering individuals and teams through environmental challenges, selecting analogs with highly salient and relevant characteristics that match space mission profiles (e.g., long versus short duration, stationkeeping versus expedition profiles) will be important.

The Expeditionary Analog

Expeditions, by definition, revolve around movement. Expeditionary analogs (e.g., oceanic, polar, desert, caving, mountaineering) include various exploratory goals that are characterized by moving from one place to another rather than inhibiting a locale. Historical exploratory expeditions typically involved long durations (i.e., months to years) characterized by significant known and unknown risks, broad goals, a high degree of situationally driven contingency decision-making, and expectations of autonomy and self-sufficiency. Modern expeditions, in contrast, are typically of short duration (i.e., two weeks to three months), utilize the advantages of technology to minimize risks (e.g., weather forecasts to take advantage of the best weather of a region and satellite communications to maintain contact), are more narrowly goal-oriented and task-focused, and involve members with specialized roles and skills. In both expeditionary scenarios, teams were/are formed around appropriate skill sets and availability and a notable lack of any attempt to screen individuals psychologically except for medical factors. Research on team functioning is often secondary to expedition goals, personal goals, schedules, and contingencies. The expedition may be intended to recreate experiences of earlier explorers, such as the Polynesian *Kon-Tiki* oceanic traverse; set records or discover new territory, e.g., discover a route to India or explore a cave system; achieve personal challenges, such as climbing mountains or skiing to the North Pole; conduct scientific research, e.g., by means of oceangoing research vessels or polar ice drilling teams; or conduct commercial exploration, such as mineral and oil exploration.[26]

---

26. Bishop, Santy, and Faulk, "Team Dynamics Analysis of the Huautla Cave Diving Expedition"; Bishop, Grobler, and Schjøll, "Relationship of Psychological and Physiological Parameters During an Arctic Ski Expedition": 261; T. Heyerdahl, *Kon-Tiki* (Chicago: Rand McNally & Company, 1950); H. R. Bernard and P. Killworth, "On the Social Structure of an Ocean Going Research

From Earth Analogs to Space: Getting There from Here

Ben Finney, Professor Emeritus in Anthropology at the University of Hawai'i and noted for his work on applying anthropological perspectives to humankind's expansion into space, has argued that from the earliest voyages to have scientific goals, "cultural" differences between scientists and seamen have led to conflict and that this inherent conflict of cultures is similarly reflected in our space program's structural differentiation between pilots and astronaut-scientists.[27] Voyages of scientific discovery began in the late 18th century, an age, Finney points out, that many have argued foreshadowed the space race of the 1960s.[28] The first exploratory voyage to include scientists as crew and mission goals with explicit scientific objectives instead of commercial goals that serendipitously collected science data was the three-year-long English expedition of the *Endeavour* to Tahiti, 1768–71, led by Captain James Cook. The on-board scientists were tasked to observe the transit of Venus across the face of the Sun to provide data needed to calculate the distance between Earth and the Sun. The success of the *Endeavour*'s expedition led to a second expedition, which sailed with a number of scientists, two astronomers, and a naturalist, an expedition that, in contrast to the first expedition, was rife with contentious relationships between the seamen and the scientists. Subsequent voyages with scientists on board were similarly plagued by conflicts between those pursuing scientific goals and those tasked with the piloting and maintenance of the ship. Historically, the English naval command eventually imposed an unofficial moratorium on the inclusion of non-naval scientists on board and pursued a policy of assigning any scientific duties to members of the crew. Not until a hundred years after Cook, in 1872, would the Royal Navy's *Challenger*, a three-masted, square-rigged, wooden vessel with a steam engine, sail around the world with six marine scientists and a crew and captain who were totally dedicated to the research.[29]

Vessel," *Social Science Research* 2 (1973): 145; H. R. Bernard and P. Killworth, "Scientist at Sea: A Case Study in Communications at Sea," Report BK-103-74, Code 452, Contract N00014-73-4-0417-0001, prepared for the Office of Naval Research (Springfield, VA: National Technical Information Service, 1974); M. M. Mallis and C. W. DeRoshia, "Circadian Rhythms, Sleep, and Performance in Space," *Aviation, Space, and Environmental Medicine* 76 (2005): B94.

27. B. Finney, "Scientists and Seamen," in *From Antarctica to Outer Space: Life in Isolation and Confinement*, ed. Harrison, Clearwater, and McKay, p. 89.

28. W. H. Goetzmann, *New Lands, New Men* (New York: Viking, 1986); J. Dunmore, *French Explorers of the Pacific*, vol. 2 (Oxford: Claredon Press, 1969).

29. E. Linklater, *The Voyage of the* Challenger (London: John Murray, 1972).

Such troubles were not limited to the English. The French followed a similar pattern, beginning in 1766 and continuing through 1800, when scientists sailed with numerous expeditions that were summarily characterized by conflict and contention between the crews and scientists.[30] Finney further notes that such complaints are found in journals of early Russian scientists, as well as American scientists on the four-year-long United States Exploring Expedition that sailed from Norfolk in 1838 with a contingent of 12 scientists.[31]

Modern development of specialized ships complete with laboratories and equipment dedicated to oceanographic research has been primarily organized and maintained by universities and oceanographic institutes. Yet even aboard these dedicated floating research vessels, conflict between the ship's crew and the scientists whom they serve has not been eliminated. A dissertation study conducted by a resident at the Scripps Institute of Oceanography during 1973 concluded that tension between the two groups was inevitable because they formed two essentially separate and distinct subcultures with different values and goals, as well as different educational backgrounds and class memberships.[32]

Finney argues that the same subcultures have become evident in the space program with the development of the role of payload specialists, who are considered visiting scientists rather than part of the elite astronaut corps. Tensions between payload specialists in pursuit of the scientific goals and the crew in pursuit of mission completion have routinely been in evidence. Finney eloquently states:

> [I]f space research were to be made as routine to the extent that ocean research now is, subcultural differences, and hence tensions, between scientist and those pilots, stationkeepers, and others whose job it will be to enable researchers to carry out their tasks in space may become critical considerations. If so, space analogues of the mechanisms that have evolved to accommodate differences

---

30. J. Dunmore, *French Explorers of the Pacific*, vol. 1 (Oxford: Clarendon Press, 1965).

31. A. von Chamisso, *Reise um die Welt mit der Romanoffischen Entdeckungs Expedition in den Jahren 1815–1818* (Berlin: Weidmann, 1856); W. Stanton, *The Great United States Exploring Expedition of 1838–1842* (Berkeley, CA: University of California Press, 1975).

32. Bernard and Killworth, "On the Social Structure of an Ocean Going Research Vessel": 145; Bernard and Killworth, "Scientist at Sea: A Case Study in Communications at Sea."

between scientists and seamen aboard oceanographic ships may have to be developed.[33]

The number and variety of expeditions examined for relevance to space is ever increasing as both modern expeditions and analyses of historical expeditions are scrutinized. An example of how examination of the records from past expeditions contributes to the current state of knowledge and provides the impetus for future studies in space can be seen in a metastudy by M. Dudley-Rowley et al. that examines written records from a sample of space missions and polar expeditions for similarities and differences in conflicts and perceptions of subjective duration of the mission. Ten missions were compared across a number of dimensions.[34] The metastudy included three space missions that represented both long- and short-duration mission profiles: Apollo 11 (1969) and Apollo 13 (1970), ranging from six to eight days apiece, and Salyut 7 (1982), which lasted over two hundred days. Four Antarctic expeditions were included: the western party field trip of the Terra Nova Expedition (1913, 48 days), an International Geophysical Year (IGY) traverse (1957–58, 88 days), the Frozen Sea expedition (1982–84, 480 days), and the International Trans-Antarctic expedition (1990, 224 days). Finally, three early Arctic expeditions were also included: the Lady Franklin Bay (1881–84, 1,080 days), Wrangel Island (1921–23, 720 days), and Dominion Explorers' (1929, 72 days). Seven factors emerged that seemed to coincide with the subjectivization of time and the differentiation of situational reality for the crews from baseline:

1. increasing distance from rescue in case of emergency (lessening chances of "returnability");
2. increasing proximity to unknown or little-understood phenomena (which could include increasing distance from Earth);
3. increasing reliance on a limited, contained environment (where a breach of environmental seals means death or where a fire inside could rapidly replace atmosphere with toxins);

---

33. Finney, "Scientists and Seamen," p. 100.
34. M. Dudley-Rowley, S. Whitney, S. Bishop, B. Caldwell, and P. D. Nolan, "Crew Size, Composition and Time: Implications for Habitat and Workplace Design in Extreme Environments" (paper presented at the SAE 30th International Conference on Environmental Systems, 10–13 July 2000).

4. increasing difficulties in communicating with Ground or Base;
5. increasing reliance on a group of companions who come to compose a microsociety as time, confinement, and distance leave the larger society behind, in a situation where innovative norms may emerge in response to the new sociophysical environment;
6. increasing autonomy from Ground's or Base's technological aid or advice; and
7. diminishing available resources needed for life and the enjoyment of life.

The missions and expeditions were ranked by prevalence of the seven factors that might correspond with the differentiation in the subjectivization of the passage of time and in the situational reality for the crews from baseline. From highest to lowest in compromising factors, the rankings fell in the following order: Lady Franklin Bay (7); Wrangel Island, Apollo 13 (6); Salyut 7 (5); Terra Nova, Apollo 11 (4); Dominion Explorers' (3); Frozen Sea (2); IGY (1); International Trans-Antarctic Expedition (0). The Lady Franklin Bay Expedition suffered 18 deaths of its complement of 25, and the rest were starving when found. The Wrangel Island expedition suffered four deaths out of its crew of five. Apollo 13 was a catastrophe that was remarkable in its recovery of the crew intact. The Salyut 7 mission, the Terra Nova western field party, and the Apollo 11 mission all had high degrees of risk. The later polar expeditions rank below these missions.

Both the space missions and the earliest polar expeditions are above or hover just below the median (3.5). Although the authors correctly note that the sample is too small to draw conclusions, the presence of similar factors in space and early polar exploration that contributed to perceptions of mission/expedition duration or of how their situational reality deviates from baseline is important to note. These results suggest that as control over their environment decreases, team members' subjective experiences of time and the situation increasingly differ from their baselines. The strong parallel between early expeditions and modern space missions lends support for historical analogs as viable substitutes for space.

Chamber Studies

Early evaluations for astronaut selection drew upon a history of sensory deprivation research initially begun by the military throughout the 1950s and 1960s to address performance concerns about two-person crews confined to armored vehicles for long durations and continued most notably through the series of studies conducted by J. P. Zubek.[35] Initially, it was believed that space would represent a significant loss of normal sensory stimulation due to isolation from people, reduction in physical stimulation, and restricted mobility. Thus, sensory deprivation chambers were argued to be good analogs for astronauts.[36] Selection procedures, therefore, included stints in dark, small, enclosed spaces for several hours to observe how potential astronauts handled the confinement and loss of perceptual cues. As Dr. Bernard Harris, the first African American to walk in space, recounts, "They put me in this little box where I couldn't move or see or hear anything. As I recall, I fell asleep after a while until the test ended."[37]

The first systematic attempts to investigate psychological adaptation factors to isolation and confinement in simulated operational environments were conducted in the 1960s and early 1970s by putting volunteers in closed rooms for several days, subjecting them to sleep deprivation and/or various levels of task demands by having them complete repetitive research tasks to evaluate various aspects of performance decrements.[38] Chamber research, as it was to become known, encompassed a variety of artificial, constructed environments whose raison d'être was control over all factors not specifically under study. Later, specially constructed confinement laboratories such as the facility at the Johns Hopkins University School of Medicine or simulators at Marshall Space Fight Center in Huntsville, Alabama; the McDonnell Douglas Corporation in Huntington Beach, California; or Ames Research Center at Moffett Field, California, housed small groups of three to six individuals in programmed environments for weeks to months of continuous residence to address a variety of space-science-related human biobehavioral issues related to group dynamics (e.g., cohesion, motivation, effects of joining and leav-

---

35. J. P. Zubek, *Sensory Deprivation: Fifteen Years of Research* (New York: Appleton-Century-Crofts, 1969); R. Honingfeld, "Group Behavior in Confinement: Review and Annotated Bibliography," Report AD0640161, prepared for the Human Engineering Lab (MD: Aberdeen Proving Ground, October 1965), p. 117.

36. B. E. Flaherty, ed., *Psychophysiological Aspects of Space Flight* (New York: Columbia University Press, 1961).

37. B. Harris, personal communication, thesis committee member (1995–96).

38. Haythorn and Altman, "Personality Factors in Isolated Environments"; Altman, "An Ecological Approach to the Functioning of Isolated and Confined Groups," p. 241.

ing established groups), performance and work productivity, communication patterns, team cooperation, and social habitability factors.

The epitome example of chamber research may be the series of four hyperbaric-chamber studies, sponsored by the European Space Agency and designed to investigate psychosocial functioning, in which groups were confined for periods lasting from 28 to 240 days.[39] Full mission protocols specifying all medical, technical, and operational parameters approximating expected living conditions of astronauts on a space station were used. The studies were intended to evaluate the efficacy of various psychosocial monitoring and assessment techniques for implementation on real space missions, as well as to investigate persistent occurrences of communication and interaction breakdowns between on-orbit teams and Mission Control anecdotally reported from space.[40] A number of opportunities and advances came from these studies, e.g., evaluating the efficacy of communication training for space teams or the opportunity to examine factors involved in an unplanned meltdown between crews precipitated by differences in cultural attitudes and norms about genders, authority, and control.[41] However, skepticism regarding the verisimilitude of studies in which discontented members can simply quit has continued to raise real concerns as to how generalizable the findings from chamber studies are to space missions.

The Middle Ground: Capsule Habitats in Extreme Unusual Environments

Occupying the middle ground between traditional expeditionary missions with moving trajectories and the artificiality of laboratory spaces designated as space

---

39. G. M. Sandal, R. Vaernes, and H. Ursin, "Interpersonal Relations During Simulated Space Missions," *Aviation, Space, and Environmental Medicine* 66 (1995): 617; G. M. Sandal, "Culture and Crew Tension During an International Space Station Simulation: Results From SFINCSS'99," *Aviation, Space, and Environmental Medicine* 75 (2004): 44.

40. Kanas, Salnitskiy, Grund, et al., "Social and Cultural Issues During Shuttle/Mir Space Missions": 647; Sandal, Vaernes, and Ursin, "Interpersonal Relations During Simulated Space Missions": 617; Gushin, Kolintchenko, Efimov, and Davies, "Psychological Evaluation and Support During EXEMSI": 283.

41. Sandal, "Culture and Crew Tension During an International Space Station Simulation": 44; D. Manzey, ed., *Space Psychology: Textbook for Basic Psychological Training of Astronauts* (Cologne, Germany: AM-BMT-DLR-98-009, ESA/EAC, 1998).

station habitats are capsule habitats, sharing the controlled, defined enclosure of the laboratory situated within an extreme unusual environment (EUE).[42] Characterized by a controlled, highly technological habitat that provides protection and life support from an environment that is harsh, dangerous, and life-threatening, capsule habitats occupy a wide range of environments. Some are true operational bases with missions in which biobehavioral research is only secondary. Others run the gamut from fundamental "tuna can" habitats with spartan support capabilities situated in locations of varying access to a full-fidelity Antarctic base constructed solely for the purposes of biobehavioral space analog research.

Submersible Habitats

Due to their high military relevance, the best-studied of capsule habitats are submarines. As an analog for space, submarines share a number of common characteristics: pressurization concerns (hyperpressurization for submarines and loss of pressurization for space), catastrophic outcomes for loss of power (e.g., the inability to return to the surface for submarines and degraded orbits for space), dependence on atmosphere revitalization and decontamination, radiation effects, and severe space restrictions. Prenuclear submarine environments were limited in the duration of submersions (72 hours), crew size (9 officers and 64 enlisted men), and deployment periods without restocking of fuel and supplies. Structurally, these short-duration mission parameters mimicked those of the early years of space, albeit with vastly larger crews. With the launch of the nuclear-powered *Nautilus* in 1954, the verisimilitude of the submersible environment as an analog for long-duration space missions was vastly improved. With the nuclear submarine, mission durations were extended to 60 to 90 days, crews were increased to 16 officers and 148 enlisted men, and resupply could be delayed for months.[43] Generalizing from submarine research to space regarding psychological and human factors related to adjustment and well-being, researchers have identified several salient issues:

---

42. Suedfeld and Steel, "The Environmental Psychology of Capsule Habitats": 227.

43. B. B. Weybrew, "Three Decades of Nuclear Submarine Research: Implications for Space and Antarctic Research," in *From Antarctica to Outer Space: Life in Isolation and Confinement*, ed. Harrison, Clearwater, and McKay, p. 103.

- atmospheric revitalization and contamination control;
- development and validation of procedures for the medical and psychological screening of recruits;
- identification of techniques for initiating and sustaining individual motivation and group morale; and
- identification of stressors, assessment of the severity of patterns of stress reactivity, and development of effective stress coping strategies.[44]

An extension of the submersible operational environment of a military submarine is the NASA Extreme Environment Mission Operations program (NEEMO) being conducted in the Aquarius underwater habitat situated off Key Largo, Florida—the only undersea research laboratory in the world. Owned by the U.S. National Oceanic and Atmospheric Administration (NOAA) and operated by the National Undersea Research Center (NURC) of the University of North Carolina at Wilmington on behalf of NOAA, Aquarius is the submerged analog to NOAA oceanic research vessels. First deployed in 1988 in the U.S. Virgin Islands and relocated to Key Largo in 1992, the underwater facility has hosted more than 80 missions and 13 crews of astronauts and space researchers since 2001. Aquarius provides a capsule habitat uniquely situated within an environment that replicates many of the closed-loop constraints of the vacuum of space, a hostile, alien environment that requires total dependency on life support; poses significant restrictions to escape or access to immediate help; and is defined by limited, confined habitable space and physical isolation. The complexity of NEEMO missions further parallels space missions in their mission architecture, with similar requirements for extensive planning, training, control, and monitoring via an external mission control entity. However, it has only been the most recent NEEMO missions in which stress, fatigue, and cognitive fitness, as well as individual and intrapersonal mood and interaction, have been the focus of study.

---

44. B. B. Weybrew, R. L. Helmreich, and N. Howard, "Psychobiological and Psychosocial Issues in Space Station Planning and Design: Inferences from Analogous Environments and Conditions" (unpublished report prepared for NASA, 1986).

From Earth Analogs to Space: Getting There from Here

Polar Stations

First and foremost, Antarctica springs to mind when polar space analogs are raised. While there are other polar bases in the Arctic and subarctic, the bulk of sustained psychological research has been conducted in Antarctica.[45] G. M. Sandal et al. conducted a recent, extensive review of the literature on psychosocial adaptation by polar work groups, expedition teams, Antarctic bases, simulation, and space crews.[46] There are 47 stations throughout the Antarctic and sub-Antarctic regions, operated by 20 different nations, with populations running from 14 to 1,100 men and women in the summer to 10 to 250 during the winter months. The base populations vary from mixed-gendered crews to male-only crews, from intact families (Chile) to unattached singletons, for assignments that last from a few months to three years.

In 1958, after the IGY (1956–57) produced the first permanent bases in Antarctica, C. S. Mullin, H. Connery, and F. Wouters conducted the first systematic psychological study of 85 men wintering over in Antarctica.[47] Their study was the first of many to identify the Antarctic fugue state later dubbed the "big-eye," characterized by pronounced absentmindedness, wandering of attention, and deterioration in situational awareness that surfaced after only a few months in isolation. The majority of subsequent studies up through the 1980s focused on the physiological changes evidenced in winter-over adaptation. Those that did address psychosocial factors tended to focus on the negative or pathological problems of psychological adjustment to Antarctic isolation and confinement, with persistent findings of depression, hostility, sleep disturbance, and impaired cognition, which quickly came to be classified as the "winter-over syndrome."[48] Sprinkled

---

45. Harrison, Clearwater, and McKay, *From Antarctica to Outer Space: Life in Isolation and Confinement*.
46. G. M. Sandal, G. R. Leon, and L. Palinkas, "Human Challenges in Polar and Space Environments," *Review Environmental Science and Biotechnology* 5, nos. 2–3 (2006), doi:10.1007/s11157-006-9000-8.
47. C. S. Mullin, H. Connery, and F. Wouters, "A Psychological-Psychiatric Study of an IGY Station in Antarctica" (report prepared for the U.S. Navy, Bureau of Medicine and Surgery, Neuropsychiatric Division, 1958).
48. E. K. E. Gunderson, "Individual Behavior in Confined or Isolated Groups," in *Man in Isolation and Confinement*, ed. Rasmussen, p. 145; Gunderson, "Psychological Studies in Antarctica," p. 115; R. Strange and W. Klein, "Emotional and Social Adjustment of Recent U.S. Winter-Over Parties in Isolated Antarctic Station," in *Polar Human Biology: The Proceedings of the*

among Antarctic research have been findings that also report positive, or salutogenic, aspects of the winter-over experience in which winter-overers have reported enhanced self-growth, positive impacts to careers, and opportunities for reflection and self-improvement.[49]

One of Antarctica's most prolific researchers, Dr. Larry Palinkas has analyzed 1,100 Americans who wintered over between 1963 and 2003 over four decades of research in Antarctica and proposed four distinct characteristics to psychosocial adaptation to isolation, confinement, and the extreme environment:

1. Adaptation follows a seasonal or cyclical pattern that seems to be associated with the altered diurnal cycle and psychological segmentation of the mission.
2. Adaptation is highly situational. Because of unique features of the station's social and physical environment and the lack of resources typically used to cope, baseline psychological measures are not as good predictors of depressed mood and performance evaluations as are concurrent psychological measures.
3. Adaptation is social. The structure of the group directly impacts individual well-being. Crews with clique structures report significantly more depression, anxiety, anger, fatigue, and confusion than crews with core-periphery structures.
4. Adaptation can also be "salutogenic," i.e., having a positive effect for individuals seeking challenging experiences in extreme environments.[50]

Palinkas found that a depressed mood was inversely associated with the severity of station physical environments—that is, the better the environment, the worse the depression—and that the winter-over experience was associated with reduced

---

SCAR/IUPS/IUBS *Symposium on Human Biology and Medicine in the Antarctic*, ed. O. G. Edholm and E. K. E. Gunderson (Chicago: Year Book Medical Publications, 1974), p. 410.

49. Mullin, "Some Psychological Aspects of Isolated Antarctic Living": 323; A. J. W. Taylor and J. T. Shurley, "Some Antarctic Troglodytes," *International Review of Applied Psychology* 20 (1971): 143–148; O. Wilson, "Human Adaptation to Life in Antarctica," in *Biogeography and Ecology in Antarctica*, ed. J. Van Meigheim, P. van Oue, and J. Schell, Monographiae Biologicae, vol. 15 (The Hague: W. Junk, 1965), p. 690; L. A. Palinkas, "Health and Performance of Antarctic Winter-Over Personnel: A Follow-Up Study," *Aviation, Space, and Environmental Medicine* 57 (1986): 954–959; D. Oliver, "Psychological Effects of Isolation and Confinement of a Winter-Over Group at McMurdo Station, Antarctica," in *From Antarctica to Outer Space: Life in Isolation and Confinement*, ed. Harrison, Clearwater, and McKay, p. 217; P. Suedfeld, "Invulnerability, Coping, Salutogenesis, Integration: Four Phases of Space Psychology," *Aviation, Space, and Environmental Medicine* 76 (2005): B61.

50. Palinkas, "On the ICE."

subsequent rates of hospital admissions.[51] He and others have speculated that the experience of adapting to the isolation and confinement, in general, improved an individual's self-efficacy and self-reliance and engendered coping skills that they used in other areas of life to buffer subsequent stress and resultant illnesses.[52]

Concordia

In 1992, France initiated plans for a new Antarctic station on the Antarctic Plateau and was later joined by Italy. In 1996, the first French-Italian team established a summer camp at Dome C to provide logistical support for the European Project for Ice Coring in Antarctica (EPICA) and begin the construction of the permanent research station. Concordia Station became operational in 2005; the first winter-over took place in February 2005 with a staff of 13. The station consists of three buildings, which are interlinked by enclosed walkways. Two large, cylindrical three-story buildings provide the station's main living and working quarters, while the third building houses technical equipment, like the electrical power plant and boiler room. The station can accommodate 16 people during the winter and 32 people during the summer season. The typical winter population consists of four technicians for the station maintenance, nine scientists or technicians for the science projects, a chief, a cook, and a medical doctor.

Dome C is one of the coldest places on Earth, with temperatures hardly rising above −25°C in summer and falling below −80°C in winter. Situated on top of the Antarctic plateau, the world's largest desert, it is extraordinarily dry and supports no animals or plants. The first summer campaign lasted 96 days, from 5 November 2005 until 8 February 2006, with 95 persons participating. The 2006 season included seven crewmembers with two medical experiments and the first two psychological experiments sponsored by the European Space Agency for which the crew acted as subjects during their stay. The two experiments investigated psychological adaptation to the environment and the process of developing group identity, issues that will also be important factors for humans traveling to Mars. For this research, the crew completed

---

51. Ibid.
52. Ibid.; Suedfeld, "Invulnerability, Coping, Salutogenesis, Integration": B61.

questionnaires at regular intervals throughout their stay. The ESA's Mistacoba experiment to profile how microbes spread and evolve in the station—an isolated and confined environment—over time started in the 2005 season, when the first crew started living at the station, and has also continued with subsequent crews. Starting from a newly built clean environment, those conducting the study took samples from fixed locations in the base as well as from crewmembers themselves.[53]

Haughton-Mars Project

One of the first of dedicated research hybrid facilities was the Haughton-Mars Project (HMP), initiated in 1996 when the National Research Council of the U.S. National Academy of Sciences and NASA Ames Research Center sponsored a postdoctoral proposal to study the Haughton Crater on Devon Island in the Canadian Arctic as a potential analog for Mars. The program has expanded from a four-member team in 1997 to a permanent habitat that hosts eight-week arctic summer field seasons with 50 to 90 participants, multiple teams, and research projects that run from instrument testing and development to biomedical and psychological evaluation. HMP routinely supports participation by NASA; the Canadian Space Agency (CSA); the Russian Institute for Space Research (IKI); various research institutions and universities in the United States, Canada, and the United Kingdom; and the U.S. Marine Corps. It has been the subject of various documentaries made by such groups as the National Geographic Society and Discovery Channel Canada.[54]

Flashline Mars Arctic Research Station

In 2000, a second dedicated research facility was deployed on Devon Island, jointly sponsored by the Haughton-Mars Project and the Mars Society: the Flashline

---

53. European Space Agency, "The Concordia Station," *http://www.concordiastation.org/* (accessed 25 May 2010); *ESA Research News*, *http://www.esa.int/esaHS/SEMBZA8A9HE_research_0.html#subhead1* (accessed 18 June 2007).

54. The Mars Institute, "NASA Haughton-Mars Project History," available at *http://www.marsonearth.org/about/history.html* (accessed 14 June 2007).

Mars Arctic Research Station (FMARS). Running concurrently with HMP, the FMARS facility was the first of four proposed analog research facilities to be built by the Mars Society, supporting smaller six-person crews for typically two- to eight-week seasons. In summer 2007, the first four-month-long FMARS mission was successfully completed with a crew of seven and a full complement of research studies covering technology, human factors, medicine, psychology, and communications.

Mars Desert Research Station

The second Mars Society station, the Mars Desert Research Station (MDRS), came online in December 2001 and is situated in the Utah desert in the American Southwest. Because of its ease of access, the American station is considered well suited as a test bed for equipment that will later be sent to more remote and unforgiving locations. For the same reason, the American station has been the focus of short-duration isolation and confinement studies since its inception. A wide range of psychological studies investigating crew factors in short-duration missions has been in place since 2002. However, beyond preliminary descriptive results presented at conferences, the small sample size of crews has necessitated waiting until enough teams have rotated through the facility to allow meta-analyses.[55] Several international teams have also used the MDRS for studies investigating comparisons between homogeneous-gendered teams, comparisons between mission teams and backup crews, and international cultural factors, among others.[56]

---

55. S. L. Bishop, S. Dawson, N. Rawat, K. Reynolds, and R. Eggins, "Expedition One: Assessing Group Dynamics in a Desert Mars Simulation" (paper presented as part of the 55th International Astronautics Conference, Vancouver, BC, 4–7 October 2004); S. L. Bishop, S. Dawson, N. Rawat, K. Reynolds, R. Eggins, and K. Bunzelek, "Assessing Teams in Mars Simulation Habitats: Lessons Learned from 2002–2004," in *Mars Analog Research*, ed. J. D. Clarke, American Astronautical Society Science and Technology Series, vol. 111 (San Diego: Univelt, 2006), p. 177.

56. S. L. Bishop, A. Sundaresan, A. Pacros, R. Patricio, and R. Annes, "A Comparison of Homogeneous Male and Female Teams in a Mars Simulation" (paper presented as part of the 56th International Astronautical Congress, Fukuoka, Japan, October 2005); S. L. Bishop, "Assessing Group Dynamics in a Mars Simulation: AustroMars Crew 48" (paper presented as part of the Mars2030: Interdisciplinary Workshop on Mars Analogue Research and AustroMars Science Workshop, University of Salzburg, Salzburg, Austria, 24–26 September 2006).

The Mars Society plans additional facilities in Iceland and Australia that will capitalize on geological features that present opportunities to practice Mars exobiology field work. The Mars Society's Mars Analog Research Station Project envisions three prime goals to be served by these habitats:[57]

- The stations will serve as effective test beds for field operations studies in preparation for human missions to Mars. They will facilitate the development and testing of key habitat design features, field exploration strategies, tools, technologies, and crew selection protocols that will enable and help optimize the productive exploration of Mars by humans. In order to achieve this goal, each station must be a realistic and adaptable habitat.
- The stations will serve as useful field research facilities at selected Mars analog sites on Earth and will help further understanding of the geology, biology, and environmental conditions on Earth and on Mars. In order to achieve this objective, each station must provide safe shelter and be an effective field laboratory.
- The stations will generate public support for sending humans to Mars. They will inform and inspire audiences around the world. As the Mars Society's flagship program, the Mars Analog Research project will serve as the foundation of a series of bold steps that will pave the way to the eventual human exploration of Mars.

CONCLUSION

The use of analogs for space is an emergent field whose very short track record examining team dynamics and psychosocial factors impacting individual and group functioning vigorously supports the real value of these environments and generalizability to space environments. Unlike laboratory studies, where the threat of real danger is usually absent, teams operating within real extreme environments have unknown situational and environmental challenges to face. Even in circumstances in which death or injury occurs, there will always be questions regarding the ability to avoid negative outcomes. While postmission analyses of behavior and performance

---

57. The Mars Society, "Mars Desert Research Station Project Goals," available at *http://www.marssociety.org/MDRS/mdrs01b.asp* (accessed 14 June 2007).

add insight into contributing factors, it is seriously doubtful whether we will ever be able to accurately predict the entire range and complexity of interaction between the human-environment factors and the human-human factors. Risk is inherent in human exploration. Even so, the value of analog experiences cannot be underestimated, regardless of whether they help us grapple with defining our levels of adequate preparation in the face of ideally predefined levels of "acceptable risk" or even "acceptable losses" (a concept familiar to those who perform military risk assessments).

One key methodological and validity issue is the added value of utilizing consistent measures across various analogs, allowing more accurate comparisons of individuals and teams across environments, including space. The necessity to validate multicultural questionnaires and methodologies that are relevant, reliable, and valid for international teams is of paramount importance as our reliance on these multinational teams will only increase in the future. To that extent, the various research endeavors in analog environments have contributed significantly to validating such assessment instruments in a variety of teams.

Findings from analogs have clearly identified three major intervention points to affect group functioning outcomes:

- Selection: the development of reliable and valid methods of choosing the best fit at both the individual and the group levels.
- Training: improving the fitness of the group by prepping skills needed for interpersonal group dynamics as well as high-functioning self-monitoring and appropriate adaptation.
- Support: taking the form of prevention first, then early, proactive intervention second. To be successful, research to date strongly suggests that the support must include the group, the family, and all external participants (e.g., Mission Control) as partners.

A large portion of the current research represents opportunities to examine team dynamics and factors that impact team function in real-world groups that have been brought together for particular purposes that have little to do with research, e.g., geological field teams. Similarly, examinations of historical sources of past expeditions will continue to inform and provide additional insight into factors that have contributed to the success or failure of previous efforts. However, we need larger, more systematic studies in which the composition of the team is one of the driving factors under investigation instead of simply an extraneous variable. Our greatest hope lies with the new research facilities now available and coming online dedicated to such research.

Chapter 4

Patterns in Crew-Initiated Photography of Earth from the ISS—Is Earth Observation a Salutogenic Experience?

Julie A. Robinson
    Office of the ISS Program Scientist
    National Aeronautics and Space Administration (NASA) Johnson Space Center (JSC)

Kelley J. Slack
    Behavioral Health and Performance Research
    Wyle Laboratories

Valerie A. Olson
    Department of Anthropology
    Rice University

Michael H. Trenchard
    Image Science and Analysis Laboratory
    Engineering and Science Contract Group (ESCG)
    NASA JSC

Kimberly J. Willis
    Image Science and Analysis Laboratory
    ESCG
    NASA JSC

Pamela J. Baskin
    Behavioral Health and Performance Research
    Wyle Laboratories

Jennifer E. Boyd
    Department of Psychiatry,
    University of California, San Francisco; and
    San Francisco Veterans Affairs Medical Center

ABSTRACT

To provide for crewmember well-being on future exploration missions, understanding coping strategies that International Space Station (ISS) crewmembers adopt to mitigate the inherent stress of long-duration confinement is important. A recent retrospective survey of flown astronauts found that the most commonly reported psychologically enriching aspects of spaceflight had to do with their perceptions of Earth. ISS crewmembers photograph Earth both volitionally and in response to requests from Crew Earth Observations (CEO) scientists. Automatically recorded data from the camera can be used to test hypotheses about factors correlated with self-initiated crewmember photography. The present study used these objective in-flight data to investigate the nature of voluntary photographic activity. We examined the distribution of photographs with respect to time, crew, and subject matter. We determined whether the frequency fluctuated in conjunction with major mission events such as vehicle dockings and extravehicular activities (EVAs, or spacewalks), relative to the norm for the relevant crew. We also examined the influence of geographic and temporal patterns on frequency of Earth photography activities. We tested the hypotheses that there would be peak photography intensity over locations of personal interest, as well as on weekends.

Of nearly 200,000 photographs taken on eight ISS expeditions, 84.5 percent were crew-initiated. Once a crewmember went to the window for a CEO request, he or she was more likely to take photographs for his or her own interest. Fewer self-initiated images were taken during and immediately preceding major station events. Crewmembers were more likely to take self-initiated images during periods when they had more free time. Analysis indicated some phasing in patterns of photography during the course of a mission, although it did not suggest that psychological functioning was lower during the third quarter of confinement (i.e., no third-quarter effect was found). Earth photography is a self-initiated positive activity of possible importance for salutogenesis (increase in well-being) of astronauts on long-duration missions. Scientific requests for photography through CEO play an important role in facilitating crew-initiated photography. Consideration should be given to developing substitute activities for crewmembers in future exploration missions where there will not be the opportunity to look at Earth, such as on long-duration transits to Mars.

*Patterns in Crew-Initiated Photography of Earth from the ISS—
Is Earth Observation a Salutogenic Experience?*

BACKGROUND
Earth Observation Throughout Human Spaceflight

John Glenn, the first U.S. astronaut in orbit, talked NASA into letting him carry a camera on *Friendship 7* on 20 February 1962.[1] On reaching orbit, Glenn told capsule communicator Alan Shepard over the radio, "Oh, that view is tremendous." Glenn proceeded to describe each of the three sunrises and sunsets he saw during the flight, and he continues to recount that experience in interviews today.[2] A number of the astronauts who have followed have verbally recounted emotional experiences related to seeing and photographing Earth, and several astronauts have documented in written form their responses to views of Earth linked to their photography activities while in space. Space Shuttle astronaut Kathryn D. Sullivan wrote in an article documented with her Earth photography, "It's hard to explain how amazing and magical this experience is. First of all, there's the astounding beauty and diversity of the planet itself, scrolling across your view at what appears to be a smooth, stately pace . . . I'm happy to report that no amount of prior study or training can fully prepare anybody for the awe and wonder this inspires."[3] Observations of familiar places on Earth can also have strong emotional connections. NASA-*Mir* astronaut Jerry Linenger recorded photographing his hometown in Michigan in his crew notebook, "Great View—Michigan + Great Lakes cloud-free—ready to go home, now!"[4]

From Apollo to the current ISS, scientists have assisted astronauts with crew-initiated and science-specific photography of Earth. All the imagery is archived in a searchable online database maintained by the descendant of the previous pro-

---

1. Jay Apt, Justin Wilkinson, and Michael Helfert, *Orbit: NASA Astronauts Photograph the Earth* (Washington, DC: National Geographic Society), pp. 11–13.

2. Bryan Ethier, "John Glenn: First American to Orbit the Earth," *American History*, October 1997, available at *http://www.historynet.com/magazines/american_history/3030096.html* (accessed 7 June 2010).

3. Kathryn D. Sullivan, "An Astronaut's View of Earth," *Update* (newsletter of the National Geographic Society's Geography Education Program) (fall 1991): 1, 12–14, available at *http://eol.jsc.nasa.gov/newsletter/uft/uft1.htm* (accessed 7 June 2010).

4. Kamlesh P. Lulla, Lev V. Dessinov, Cynthia A. Evans, Patricia W. Dickerson, and Julie A. Robinson, *Dynamic Earth Environments: Remote Sensing Observations from Shuttle–Mir Missions* (New York: John Wiley & Sons, 2000).

grams on the International Space Station, CEO, which provided statistics summarized here. Over 2,500 photographs of Earth were taken by Mercury and Gemini astronauts. Apollo astronauts photographed both Earth and Moon views, with over 11,000 photographs taken, and have been credited with initiating the interest in Earth observations from space.[5] Handheld photography of Earth by astronauts on *Skylab* accompanied the extensive imagery obtained by an automated multispectral camera system.[6] Over the three *Skylab* missions, crewmembers took around 2,400 images of Earth, and the automated camera systems an additional 38,000 photographs with specialized films.

Building from this experience and the growing interest in Earth observations from space, a program called the Space Shuttle Earth Observations Project (SSEOP) was established in 1982 to support the acquisition and scientific use of Earth photography from Space Shuttle flights. Located at the center of astronaut training, Johnson Space Center, SSEOP scientists were assigned to each Shuttle crew. Astronauts were trained in geology, geography, meteorology, oceanography, and environmental change for a total of approximately 12 instructional hours prior to flight. Also before flight, about 20 to 30 sites were chosen for the crew to photograph while on orbit. The mission-specific sites were chosen from a list of previously identified environmentally dynamic terrestrial areas visible from the Space Shuttle. Each crew was given a preflight manual consisting of their unique sites that included photographs and scientific information. The decision on when to take photographs was at the astronauts' discretion. A list of targets was sent to the Shuttle crew on a daily basis during the flight. The main camera used for Earth observation was the 70-millimeter Hasselblad with the 50-, 100-, 110-, and 250-millimeter lenses commonly used, and both color and infrared film was made available per crew preference.[7] After each flight, the Earth-viewing film was cataloged and entered into a database. Paper catalogs were also mailed to a subscriber list of interested scientists

---

5. Paul D. Lowman, Jr., "Landsat and Apollo: The Forgotten Legacy," *Photogrammetric Engineering and Remote Sensing* 65 (1999): 1143–1147.

6. NASA, *Skylab Earth Resources Data Catalog*, JSC-09016 (Houston, TX: Johnson Space Center, 1974); V. R. Wilmarth, J. L. Kaltenbach, and W. B. Lenoir, eds., *Skylab Explores the Earth* (Washington, DC: NASA SP-380, 1977), pp. 1–35.

7. Julie A. Robinson, David L. Amsbury, Donn A. Liddle, and Cynthia A. Evans, "Astronaut-Acquired Orbital Photographs as Digital Data for Remote Sensing: Spatial Resolution," *International Journal of Remote Sensing* 23 (2002): 4403–4438.

*Patterns in Crew-Initiated Photography of Earth from the ISS—
Is Earth Observation a Salutogenic Experience?*

and educational and government users. To date, Shuttle crewmembers have captured over 287,000 images of Earth.

From March 1996 through June 1998, the scientists of SSEOP supported Earth photography by crewmembers spending longer durations in space as part of the NASA-*Mir* program. U.S. investigators collaborated with the Institute of Geography, Russian Academy of Sciences, in developing Earth observation objectives for astronauts on board *Mir*.[8] The documentation of dynamic environmental changes on Earth's surface was a primary objective for both SSEOP and the Russian Institute of Geography. Another objective was to develop scientific approaches and procedures that could later be applied to the same kinds of dynamic observations from the ISS. With the advent of Shuttle-*Mir* and the ISS, the focus of SSEOP changed from short-term observation to long-term observation.

The cameras used on Shuttle-*Mir* were the same as on the Shuttle, with the 70-millimeter Hasselblad (film) as the main camera, but the Nikon F3 35-millimeter camera was also available. A joint list of sites was chosen by U.S. and Russian scientists for Shuttle-*Mir*. Earth observation target sites were sent to the Shuttle-*Mir* crews weekly. Training was modified from the typical Shuttle briefings to enable the Shuttle-*Mir* crews to document unanticipated dynamic events as well as targets of opportunity that would be encountered more often on long-term missions. Another benefit of long-term Earth observing missions was the ability to document seasonal change and long-term climatic effects. Approximately 22,500 photographs were taken during the seven Shuttle-*Mir* missions.

Crew Earth Observations began as a formal ISS research activity ("payload") on the first mission, Expedition 1, in October 2000. Training for ISS crews evolved from experiences gained in the Shuttle and Shuttle-*Mir* programs. Rather than discipline-specific training, ISS crews were trained on science topics such as coral reefs, global urban systems, deltas, and glaciers. The emphasis was more on observing Earth as a system than on documenting independent events. An overall science plan tied together the target sites and crew training and is still used and updated by increment for ISS crews today. Due to the extensive training ISS astronauts receive regarding all aspects of their missions, CEO training is limited to 4 hours. Typically this training occurs during the early part of the training cycle. Since an

---

8. Lulla, Dessinov, Evans, Dickerson, and Robinson, *Dynamic Earth Environments*.

ISS mission is longer than a Shuttle mission, the number of targets per increment varies from approximately 140 to 160 sites, and they are updated with the change of each ISS increment.

The digital camera, a Kodak 460 DSC, was introduced on STS-73; however, the Hasselblad film camera remained the favorite of the Shuttle crews, most likely because of their experience with that camera. Improvements in digital technology coincided with the change in focus of the Shuttle program to the assembly of the International Space Station. Following the Space Shuttle *Columbia* accident in 2003, NASA's support of Earth observations by crewmembers has been focused on the ISS. Although SSEOP was dissolved, individual Shuttle crewmembers on missions to the ISS could still use the on-board cameras to take images of Earth, but without scientific support.

Earth Observation in Human Spaceflight Today

The digital camera was favored by ISS crews over the film cameras because it allowed them to review their imagery while on orbit. The immediate review of their imagery enabled the crews to view and improve their photographic techniques. Digital images could also be down-linked to the CEO scientists for review, and the scientists in turn could provide feedback to the crew. The issue of film versus digital cameras was settled in 2003 when mission length was extended to about six months. The extension of crew time on orbit made film more susceptible to radiation "fogging." While digital cameras are not immune to radiation, they are better able to cope with longer exposures to the space environment, and eliminating the need to return film to Earth was also an important improvement.

With the use of the 400-millimeter lens and 2× extender available for the digital camera, ISS crews have been able to document dynamic events at a higher resolution than was possible from the Shuttle with the 250-millimeter lens.[9] The 400- and 800- millimeter lens options are clearly the favorites of ISS crews. An additional benefit of the camera is the automatic logging of the time as well as

---

9. Julie A. Robinson and Cynthia A. Evans, "Space Station Allows Remote Sensing of Earth to Within Six Meters," *Eos, Transactions, American Geophysical Union* 83 (2002): 185, 188.

the date the image was acquired, along with other camera settings. Currently, the Kodak 760 DSC is used for CEO; however, this camera was upgraded with the higher resolution Nikon D2x in the latter part of 2008.

In addition to watching Earth, ISS crewmembers photograph Earth through the windows of the ISS and are able to share those images with the world. The CEO activity provides a venue to transmit requests for photographs of areas of scientific or public interest to the astronauts each day and to distribute the acquired photographs to scientists and the public. Crewmembers take photographs of the targets during their free or unscheduled time; Earth photography is never a scheduled crew activity. A list of candidate targets is sent to them on a daily basis, and crewmembers can make attempts to photograph those targets, choose to take no images, or, on their own initiative, photograph Earth at any time. These self-initiated images would seem to be of special importance to crewmembers since the taking of these images is purely volitional. Whether requested by scientists or self-initiated, images of Earth taken from the ISS are identified and distributed via the *Gateway to Astronaut Photography of Earth* Web site.[10]

Earth Observation and Behavioral Health in Human Spaceflight

While NASA has always engaged in space exploration research, *The Vision for Space Exploration* and subsequent definitions of specific exploration mission architectures have required a much more focused use of the ISS.[11] In particular, the ISS is to be used for research on human health on long-duration space missions, as well as for technology development and testing.[12] Behavioral health and performance has been identified as a discipline with additional research needs requiring the ISS.[13]

---

10. NASA, *Gateway to Astronaut Photography of Earth* Web site, http://eol.jsc.nasa.gov.

11. NASA, *The Vision for Space Exploration* (Washington, DC: NASA NP-2004-01-334-HQ, 2004), pp. 15–17.

12. NASA, "The NASA Research and Utilization Plan for the International Space Station (ISS), A Report to the Committee on Science of the United States House of Representatives and the Committee on Commerce, Science, and Transportation of the United States Senate, NASA Headquarters" (Washington, DC: NASA Headquarters, 2006), pp. 1–20.

13. NASA, *Bioastronautics Roadmap: A Risk Reduction Strategy for Human Space Exploration* (Houston, TX: NASA Johnson Space Center SP-2004-6113, 2005), p. 5; John R. Ball and

Maximizing psychological well-being and performance of the crew, while in a confined space with interpersonal interactions limited to a small number of people, is important for the success of ongoing ISS missions. Knowledge about behavioral health gained from ISS missions is also important for the success of future missions to a lunar base and provides key data for a four- to six-month Mars transit. A particular concern is maintaining crew psychological well-being for the duration of a round-trip mission to Mars that could last as long as three years.[14]

Positive (or "salutogenic") experiences while in space may promote psychological well-being by enhancing personal growth and may be important for offsetting the challenges of living and working in a confined and isolated environment.[15] In a survey of flown astronauts aimed at identifying the positive or salutogenic effects of spaceflight, Eva Ihle and colleagues identified positive changes in perceptions of Earth as the most important change experienced by astronauts.[16]

If viewing Earth is an important component of positive experiences in spaceflight, then having Earth "out of view" may be an important challenge for crews going to Mars because it could increase the sense of isolation.[17] To the extent that observing Earth is a positive experience for ISS crewmembers, replacement activities or new psychological countermeasures may be needed to ensure the well-being of crewmembers on a Mars mission.

Charles H. Evans, Jr., eds., *Safe Passage: Astronaut Care for Exploration Missions* (Washington, DC: Committee on Creating a Vision for Space Medicine During Travel Beyond Earth Orbit, Board on Health Sciences Policy, National Institute of Medicine, National Academy Press, 2001), pp. 136–171.

14. NASA, *Bioastronautics Roadmap*, p. 12.

15. Peter Suedfeld, "Applying Positive Psychology in the Study of Extreme Environments," *Journal of Human Performance in Extreme Environments* 6 (2001): 21–25; Peter Suedfeld and Tara Weiszbeck, "The Impact of Outer Space on Inner Space," *Aviation, Space, and Environmental Medicine* 75, no. 7, supplement (2004): C6–C9.

16. Eva C. Ihle, Jennifer Boyd Ritsher, and Nick Kanas, "Positive Psychological Outcomes of Spaceflight: An Empirical Study," *Aviation, Space, and Environmental Medicine* 77 (2006): 93–101.

17. Nick Kanas and Dietrich Manzey, *Space Psychology and Psychiatry* (Dordrecht, Netherlands: Kluwer Academic Publishers, 2003), p. 186.

*Patterns in Crew-Initiated Photography of Earth from the ISS—
Is Earth Observation a Salutogenic Experience?*

Objectives

In this paper, we mine the dataset of Earth observation photography to see whether additional information could be gleaned about the importance to crewmembers of the positive experience of viewing Earth. Our first objective was to quantify the extent to which photography of Earth was self-initiated. A second objective was to identify patterns in photography, or conditions under which crewmembers were more likely to take self-initiated images. From this we hoped to gain quantitative (although correlative) insight into whether Earth observation activities are important to long-duration crewmembers on the ISS and use this to infer whether Earth observation activities might play a role in maintaining the psychological well-being of at least some of these crewmembers.

Hypotheses

Prior to analyzing the photographic incidence data, we generated the following hypotheses:

Hypothesis 1: Fewer self-initiated images are expected to be taken during periods of, and preparation for, extraordinary activities. Daily activities on the Station can be very crudely dichotomized into regular daily activities and extraordinary activities. Extraordinary activities include EVAs as well as docking and undocking (i.e., of Space Shuttle, Soyuz, and Progress spacecraft). Further, these extraordinary activities require substantial focus and preparation leading up to the event. These extraordinary activities generally consume more time than regular daily activities, leaving less time for volitional activities such as taking images. In the mission timelines, extensive EVA and docking preparation ramps up prior to an event, with restrictions on the ability to schedule other, noncritical activities beginning one week prior to the EVA, so we considered one week prior as our preparation period.

Hypothesis 2: More self-initiated images are expected to be taken during weekends or other light-duty times. Typically, crewmembers have fewer set tasks to accomplish on weekends, so they have increased periods of time in which they can choose their activities. Given the volitional nature of self-initiated images coupled with the enjoyment crews have stated that they receive from viewing

Earth, we expected crewmembers to take more Earth photographs during periods of decreased workload.

Hypothesis 3: More self-initiated images are expected to be taken of geographic areas of personal interest to crewmembers. Past crews have placed great importance on viewing Earth,[18] and most Shuttle and ISS crewmembers have requested support in photographing their hometowns and other places of personal interest. If such interest provides an indirect linkage of crewmembers in space to the people and place they have left behind, the photographing of places that hold special meaning for crewmembers, such as their childhood home or their alma mater, might be expected to be of particular relevance.

Hypothesis 4a: Phasing occurs such that differing numbers of self-initiated images will be taken over the course of a mission. Hypothesis 4b: During the third quarter of the mission, increased numbers of self-initiated images will be taken. Previous research, both in space and in analog environments such as the Antarctic, has found mixed results regarding the existence of either phasing or a third-quarter effect.[19] The term *phasing* suggests that isolated individuals experience a cycle of ups and downs in psychological well-being during their time in confinement. While the term *phasing* is more general, the term *third-quarter effect* specifically refers to a period of lowered psychological well-being during the third quarter of an extended confinement. Thus, we looked for several possible temporal patterns in the incidence of self-initiated photography.

---

18. Ihle et al., "Positive Psychological Outcomes of Spaceflight": 93–101.

19. Robert B. Bechtel and Amy Berning, "The Third-Quarter Phenomenon: Do People Experience Discomfort After Stress Has Passed?" in *From Antarctica to Outer Space: Life in Isolation and Confinement*, ed. Albert A. Harrison, Yvonne A. Clearwater, and Christopher P. McKay (New York: Springer Verlag, 1991), pp. 261–266; Mary M. Connors, Albert A. Harrison, and Faren R. Akins, *Living Aloft: Human Requirements for Extended Spaceflight* (Washington, DC: NASA SP-483, 1985); Nick Kanas, Daniel S. Weiss, and Charles R. Marmar, "Crew Member Interactions During a Mir Space Station Simulation," *Aviation, Space, and Environmental Medicine* 67 (1996), 969–975; Gro M. Sandal, "Coping in Antarctica: Is It Possible to Generalize Results Across Settings?" *Aviation, Space, and Environmental Medicine* 71, no. 9, supplement (2000): A37–A43; Jack W. Stuster, Claude Bachelard, and Peter Suedfeld, "The Relative Importance of Behavioral Issues During Long-Duration ICE Missions," *Aviation, Space, and Environmental Medicine* 71, no. 9, supplement (2000): A17–A25.

METHODS
Participants

Images taken by up to 19 ISS crewmembers, beginning with ISS Expedition 4 (December 2001, when the full capability of the digital camera began to be used) and continuing through Expedition 11 (October 2005), were included in this study. Ten were astronauts with NASA, and nine were Russian cosmonauts. The expeditions consisted of three crewmembers through Expedition 6, when the number of crewmembers on the Station dropped to two, one Russian and one American. Gender of the crew for Expeditions 4 through 11 was predominantly male with only one female astronaut. It is not known whether every individual on board the ISS actually used the camera, nor which individuals took which images.

Data and Analyses

Digital photographs are taken on orbit and downlinked to the ground during the course of the mission. These are separated by content (Earth, hardware, people). All Earth images become part of the Database of Astronaut Photography of Earth, which was used for these analyses and is available online.[20]

We analyzed the Earth photography patterns using the digital data recorded on the back of the digital cameras used on the ISS. The cameras automatically record the date and time when the photograph was taken, as well as specific photographic parameters. The data do not identify the individuals using the camera, as any crewmember may pick up any camera to take pictures, and individuals often stop briefly at a window to take pictures throughout the day. Crews are cross-trained in the use of the imagery equipment. Some crews share the responsibility of taking images of Earth; in other crews, one member might have more interest and thus be the primary photographer. Regardless, crewmembers report photographing areas known to be of interest to fellow crewmembers.

---

20. NASA, *Gateway to Astronaut Photography of Earth* Web site, http://eol.jsc.nasa.gov (accessed 9 December 2010).

Additional datasets compiled for use in analyses were 1) lists of areas of known geographic interest to crews based on publicly released biographical information, 2) orbital track parameters to relate images taken to the log of scientific requests sent to the crew, and 3) records of on-orbit activities to determine the incidence of EVAs, the docking of visiting vehicles, and days of light duty/holidays. We used the orbit tracks and message logs to identify which photographs were in response to CEO requests and which were self-initiated by the crew. Occasionally, battery changes and camera resets were conducted on orbit without resetting the date and time on the camera. Because of this, not all camera time stamps were accurate. We screened those data for inaccuracies (such as an incorrect year for a specific expedition), and these records were eliminated from the analyses.

For each day, we determined the number of images of Earth that were self-initiated, were of areas of known geographic interest to any member of that crew, were in response to a scientific request, and used the 800-millimeter (high-magnification) lens setup. The use of the 800-millimeter lens was tracked because it represents a significant skill that requires much effort to achieve the best results, and the resulting images provide the most detail (up to 6-meter spatial resolution). The crewmembers must practice tracking the motion of Earth beneath the ISS using the camera equipped with the 800-millimeter lens and learn how to focus properly through the lens.[21] Although this was not one of our original hypotheses, we realized that use of the 800-millimeter lens could be an indicator of crew interest in Earth photography as a challenging, self-motivated hobby.

In general terms, the analyses looked for relationships between self-initiated image-taking and when the images were taken, as well as between self-initiated image-taking and the geographic location of those images. For the benefit of statistically minded readers, hypotheses 1 and 2 were addressed by examining zero order correlations and using general linear models in a statistical analysis package (GLIMMIX [generalized linear mixed models] procedure in SAS). This procedure fits generalized linear mixed models to the data and allows for normally distributed (Gaussian) random effects.[22] Hypothesis 3 was tested using a related procedure that could incorporate categorical data into the model (GENMOD procedure for gen-

---

21. Robinson and Evans, "Space Station Allows Remote Sensing of Earth to Within Six Meters": 185.
22. SAS, *The GLIMMIX Procedure* (Cary, NC: SAS Institute Inc., June 2006), p. 5, available at *http://support.sas.com/rnd/app/papers/glimmix.pdf* (accessed 5 May 2006).

eralized linear models in SAS). Hypothesis 4a was tested using regression, while general linear model repeated measures analysis was used for hypothesis 4b.

RESULTS

From December 2001 (Expedition 4) through October 2005 (Expedition 11), crewmembers took 144,180 images that had accurate time and date data automatically recorded by the camera. Of time-stamped photographs, 84.5 percent were crew-initiated and not in response to CEO requests.

Comparison of Variables

These comparisons were made by examining the degree to which the variables are related (correlation), the average for each variable (mean), and the degree to which the values for a variable differ from its average (standard deviation). See table 1 for all measures included in the study. For subsequent analyses, we considered only self-initiated images and excluded images taken in response to CEO requests.

The correlations presented in table 1 provide a preliminary examination of the data rather than a formal test of the hypotheses. When conducting statistical analyses, correlations typically are examined first, and then a priori hypotheses are tested using more robust statistical approaches. Based on the correlations in table 1, the following inferences can be made.

A crewmember with a camera in hand was more likely to take self-initiated photos in addition to the requested images (self-initiated images were correlated with requested images—$r = .36, p < .01$). Also, 800 millimeters was the focal length more frequently chosen when crewmembers took images of their own choice, even though taking images at 800 millimeters was more challenging (self-initiated images were correlated with images taken at an 800-millimeter focal length—$r = .41, p < .01$). Further, crewmembers also were more likely to take self-initiated images of geographic areas of Earth that were of personal interest to them ($r = .25, p < .01$). A crew containing a member, for example, whose childhood home was in a small town in Illinois, would be more likely to take images of that area than of areas not holding personal significance for any member of that crew.

Table 1. Means, standard deviations, and correlations across all missions. Each parameter is measured on a daily basis across all expeditions combined.

	Daily Number of:	Mean	Std. Dev.	1	2	3	4	5	6
1	Total images taken	102.3	119.1	—					
2	Self-initiated images taken	86.4	107.5	.98**	—				
3	Images of geographic interest	1.6	5.1	.25**	.25**	—			
4	Requested images taken	15.9	25.3	.54**	.36**	.10**	—		
5	Images taken with 800-mm	17.8	34.4	.41**	.41**	.15**	.19**	—	
	Proportion of Days:	Mean	Std. Dev.	1	2	3	4	5	6
6	Higher availability to take images	.3	.4	.06*	.07**	-.01	-.03	.07**	—

*   Correlation is significant at the .05 level (2-tailed).
** Correlation is significant at the .01 level (2-tailed).

Crews were more likely to take self-initiated images on weekends (self-initiated images correlated with weekends—$r = .07, p < .01$) However, contrary to expectations, neither activity nor holiday was related to self-initiated images (due to space limitations, the variables of weekends, activities, and holidays are not included in table 1). It could be that crewmembers did not necessarily have more time available on holidays or that self-initiated images are more tightly linked to scientist-requested approaches to the window. To address the possibility that holidays and planned activities were not indicative of whether crewmembers had time available to take self-initiated images, the variables of holiday, weekend, and activity were combined to create a measure of general availability to take images. Using this new measure of general availability, we found that crewmembers were more likely to take self-initiated images when they had time available (general availability was correlated with self-initiated images—$r = .07, p < .01$).

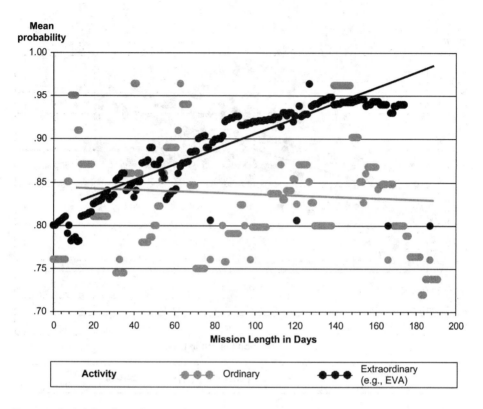

Figure 1. Probability that self-initiated images were taken, using activity as a predictor across mission.

Tests of Hypotheses

Hypothesis 1: Activity. Hypothesis 1 stated that fewer self-initiated images would be taken one week prior to and during extraordinary mission events, such as EVAs and dockings. The data analysis supported this hypothesis. Crewmembers were less likely to take self-initiated images while preparing for and during these mission events than during regular daily station activities. To state in statistical terms, conditional upon the degree of variability that could not be described by the model (the random effect of the intercept) and controlling for time (days) on the ISS, the type of activity on the Station predicted whether self-initiated images were taken ($t = -2.50$, $p < .01$). Further, crewmembers were more likely to take self-initiated images on days of regular Station activities as the mission progressed. In other words, the longer crewmembers had been on orbit, the more likely they were to take self-initiated images (the effect of activity also was different over time—$t = -4.65$, $p < .01$) (see figure 1).

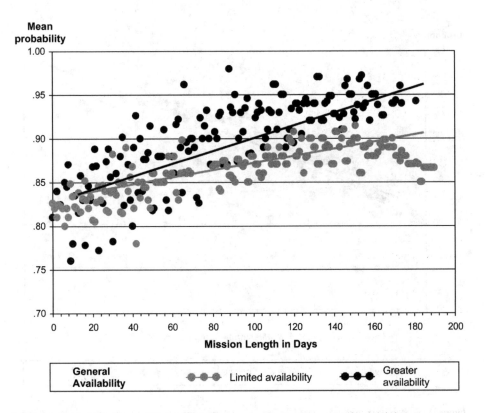

Figure 2. Probability that self-initiated images were taken, using general availability as a predictor across mission.

Extraordinary mission events, along with the period spent in orbit, influenced whether or not images were taken each day but did not influence the number of images taken on a given day. To use other words, although general activity predicted whether or not any photographic activity occurred, it did not predict the *number* of self-initiated images taken during regular or extraordinary mission events ($t = -.77$, not significant [ns]).

Hypothesis 2: Weekends. According to hypothesis 2, more self-initiated images would be taken on weekends. This hypothesis was not supported ($t = .65$, ns), perhaps because life on the Station does not always allow weekends off. This finding, in apparent conflict with the preliminary finding that self-initiated images taken were correlated with weekends, is due to hypothesis 2's being tested with more robust statistical methods than those of correlation alone.

*Patterns in Crew-Initiated Photography of Earth from the ISS—
Is Earth Observation a Salutogenic Experience?*

In response to this supposition, the composite variable of days available, or general availability, was used to reflect workload on the ISS more accurately. This composite took into account whether activity on the Station was extraordinary, whether it was a weekend, and whether a particular day was considered an off or partial-duty day or a regular-duty day. General availability was thus a more realistic representation of days with crewmember free time.

This post hoc hypothesis was supported. Regarding the taking of images, more self-initiated images were taken when crewmember schedules allowed (see figure 2). Crewmembers were less likely to take self-initiated images on a Saturday, for example, if they were also preparing for an EVA. In more statistical terms, general availability was associated with whether self-initiated images were taken ($t = 4.37$, $p < .01$), conditional upon the intercept and after controlling for time on the Station. Additionally, later in the mission, crewmembers became even more likely to take self-initiated images when their schedules allowed, indicating that general availability predicted differentially as the mission progressed ($t = 2.26, p = .02$).

Regarding the *number* of images taken, the longer crewmembers had available time, the more images they took. In more statistical terms, general availability predicted the number of self-initiated images taken ($t = 2.10, p = .02$) after controlling for the intercept and time on the Station. Further, crewmembers took just as many images when they had time available at the beginning of the mission as at the end of the mission. In other words, general availability did not differentially predict over the length of the mission ($t = -1.22$, ns).

Hypothesis 3: Geographic interest. Hypothesis 3 stated that crewmembers would take more self-initiated images of areas that were of personal interest to them.[23] This hypothesis was not supported by the data (χ^2 (df = 1) = 1.14, ns). Crewmembers were no more likely to take self-initiated images of geographic areas of personal interest to them than they were to take any other self-initiated images. The fidelity of the measure of areas of personal interest is questionable, though. The geographic areas of interest were determined by our survey methods rather than by direct reports from crewmembers themselves.

---

23. Due to limitations in the GLIMMIX procedure, a model fitting geographic interest could not converge. As an alternative, the GENMOD procedure was used. Given that GLIMMIX uses the GENMOD procedure to fit generalized linear models when random effects are absent, this change in statistical procedure is not significant.

Psychology of Space Exploration

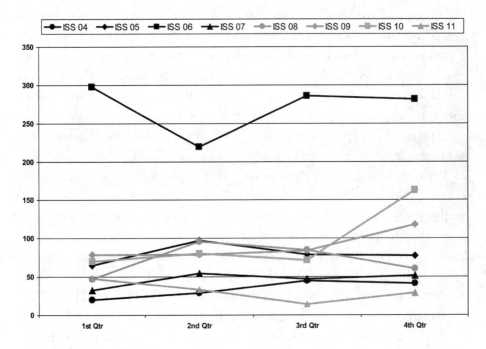

Figure 3. Quarterly estimated marginal mean number of self-initiated images taken by mission.

Hypotheses 4a and 4b: Phasing/third-quarter effects. Hypothesis 4a stated that phasing would occur such that self-initiated images would be taken differentially across the mission. This hypothesis was supported in that the number of self-initiated images taken is not consistent across the duration of the mission. Statistically, the quadratic term of the regression equation used to assess curvilinearity in the data was significant ($R^2 = .02$, $p < .01$). The temporal effects identified in the tests of hypothesis 1 and, to a lesser extent, hypothesis 2 lend further support to hypothesis 4a. Conditional upon the type of activity, the longer the crewmembers were on the Station, the more likely they were to take self-initiated images. In more statistical terms, when using type of activity as a predictor, time on the Station was a significant predictor of whether self-initiated images would be taken ($t = 3.16$, $p < .01$).

Hypothesis 4b stated that greater numbers of self-initiated images would be taken during the third quarter of the mission. This hypothesis was not supported; figure 3 effectively illustrates the lack of a third-quarter effect. The statistical method used to test hypothesis 4b was general linear model repeated measures.

*Patterns in Crew-Initiated Photography of Earth from the ISS—
Is Earth Observation a Salutogenic Experience?*

Figure 4. Example of a crew-initiated image of São Paulo, Brazil, at night. While staying on the ISS, astronaut Don Pettit assembled a homemade tracking system to photograph long-exposure images under low light conditions. *(Image number ISS006-E-44689, 12 April 2003)*

DISCUSSION

In this study, we made novel use of data available on the patterns of Earth photography by crews aboard the International Space Station. Although the data are observational, they allow additional insights into the role that observing Earth plays in the daily lives of crews in orbit. Perhaps the most important observation is the importance that photography of Earth has for at least some crewmembers, as evidenced by the degree to which it is self-initiated (84.5 percent of 144,180 photographs) and conducted as a leisure activity (for an example, see figure 4). As of mid-2006, active photography of Earth has continued, with a total over 250,000 images taken through Expedition 13.

Earth photography also offers several venues for personal accomplishments for ISS crewmembers. For example, the use of motion tracking with the 800-millimeter lens combination to achieve high-quality, high-magnification images of Earth is a challenge that some crewmembers have relished. Anecdotally, some crewmembers

Psychology of Space Exploration

Figure 5. Example of a crew-initiated image using the 800-millimeter lens combination. This view shows a portion of an image of the Golden Gate Bridge, San Francisco, California, taken by astronaut Jeff Williams from the ISS during Expedition 13. Expedition 13 held the record for the largest number of Earth images taken during an expedition (74,266 Earth images taken) until it was surpassed when Jeff Williams was again on the ISS (this time with Soichi Noguchi and T. J. Creamer) during Expedition 22 (88,779 images), November 2009–March 2010. *(Image number ISS013-E-65111, 6 August 2006)*

*Patterns in Crew-Initiated Photography of Earth from the ISS—
Is Earth Observation a Salutogenic Experience?*

Figure 6. View of Mount McKinley (Denali), Alaska, taken from the ISS using a powerful 800-millimeter lens to photograph this peak while the spacecraft was over the Gulf of Alaska, 800 miles to the south of the mountain. Cosmonaut Sergei Krikalev has assembled a collection of these views of major world mountain peaks during his stays on the ISS during Expeditions 1 and 11. *(Image number ISS011-E-11806, 14 August 2005)*

do not try to use the combination; others make it a personal challenge, and several crewmembers have become expert at its use, taking many thousands of such images during their mission (for an example, see figure 5). We cannot distinguish whether the most important element of this activity is taking and sharing these sometimes astounding images of Earth, the challenge of learning/perfecting a new skill, or a combination (for an example, see figure 6).

The correlation between scientific requests and self-initiated photography might reflect the practicality of a crewmember who continues to photograph Earth once he/she has a camera in hand. However, it is also suggestive of the importance of the scientific basis and public use of photographs in making the activity worthwhile for the crews. The scientific support from Crew Earth Observations enables self-initiated photography of Earth by providing opportunities and reminders to go to the window with a camera. It is likely that the image identification and Web distribution completed by CEO adds value to the self-initiated photography by ensuring that it can be shared with the world—but confirming this hypothesis would required a more structured survey of crewmembers. The database structure of CEO also adds personal value to crewmembers, making it easier for them to search for images they have taken for their own use.

From our analyses for hypotheses 1 and 2, self-initiated images were less likely to be taken when workload prevented it—and since 84.5 percent of images taken are self-initiated, Earth photography is clearly a leisure activity. However, crews are more likely to take self-initiated images as the mission progresses—perhaps due to acclimation and familiarity with life and duties on the Station or a growing realization that their time in space, and thus their ability to photograph Earth from space, is limited. This trend over the duration of the mission was the only mission phasing observed. A more careful examination of figure 3 suggests that the phasing effect might be due more to individual differences pertaining to specific missions or perhaps to an increasing competency with the photographic equipment. It is not clear to what extent this phasing effect might reflect differences in mission profiles or characteristics of the particular crewmembers assigned to the particular missions.

Future Research and Applications

In spite of the importance of behavioral health and performance for the success of human spaceflight missions, relatively few studies have been done on the ISS to date.[24] This analysis of data collected for other purposes serves as an example of mining data collected as part of ISS operations to increase knowledge. Such analyses can inform surveys for future data collection and influence future behavioral studies on the ISS. Given these observations, future studies should consider crew motivations in photographing Earth. Psychological component testing could reveal whether taking more self-initiated images is associated with increased crew psychological well-being.

---

24. Cynthia A. Evans, Julie A. Robinson, Judy Tate-Brown, Tracy Thumm, Jessica Crespo-Richey, David Baumann, and Jennifer Rhatigan, *International Space Station Science Research Accomplishments During the Assembly Years: An Analysis of Results from 2000–2008* (Houston, TX: NASA Johnson Space Center, NASA/TP-2009-213146–REVISION A, 2009), pp. 110–159.

Patterns in Crew-Initiated Photography of Earth from the ISS— Is Earth Observation a Salutogenic Experience?

The importance of viewing Earth as reported in crew surveys, crewmember anecdotes to CEO personnel (and the published quotations in the introduction), and the patterns in photography of Earth reported in our analysis all point to a positive psychological role on the part of these activities. Conclusions from this type of correlative data mining are only suggestive of the importance of Earth observation for crewmembers. Future studies of behavioral health should consider quantitative assessment of salutogenic effects of leisure activities such as Earth observations.

As we begin to plan for interplanetary missions, it is important to consider what types of activities could be substituted. Perhaps the crewmembers best suited to a Mars transit are those individuals who can get a boost to psychological well-being from scientific observations and astronomical imaging. Replacements for the challenge of mastering 800-millimeter photography could also be identified. As humans head beyond low-Earth orbit, crewmembers looking at Earth will only see a pale-blue dot, and then, someday in the far future, they will be too far away to view Earth at all.

ACKNOWLEDGMENTS

We would like to thank Edna Fiedler and Frank Carpenter for their encouragement to pursue these analyses. Chuck Green advised us on the statistical analyses and helped with access to SAS procedures. We also thank Cindy Evans, Edna Fiedler, and Nick Kanas for their reviews and suggestions for improvement.

Chapter 5

Managing Negative Interactions in Space Crews: The Role of Simulator Research

Harvey Wichman
    Aerospace Psychology Laboratory
    Claremont McKenna College and Claremont Graduate University

ABSTRACT

This chapter argues that a watershed period has been reached in the history of spaceflight that requires a "paradigm shift" in the way spacecraft are designed and people are selected and trained for spaceflight. In the beginning, space programs had minimal spacecraft, and flights were of short duration. Heroic human specimens were then recruited and extensively trained to perform in these machines no matter how difficult or uncomfortable it was. Spacecraft technology is now sufficiently sophisticated to design spacecraft to be much more accommodating to human occupants. The historical timing of this shift in thinking is heralded by the coming together of sophisticated space technology, the rise of space tourism, and the desire for spaceflights of greater duration than brief sorties into Earth orbit.

Simulator technology has developed in step with spacecraft technology. However, simulators are used primarily in training. The chapter concludes with an illustration of how simulators can be used as behavioral research laboratories. A study conducted for the McDonnell Douglas Aerospace Corporation is presented; in it, a spaceflight simulator was used to explore both applied and theoretical questions with a diverse group of civilian passengers in a simulated 45-hour orbital spaceflight.

INTRODUCTION

In the 47 years since Yuri Gagarin became the first person in space and the first person to orbit Earth, several hundred cosmonauts and astronauts have successfully flown in space. Clearly, there is no longer any doubt that people can live and work successfully in space in Earth orbit. This ability has been demonstrated in spacecraft as tiny as the Mercury capsules, in Space Shuttles, and in various (and much more spacious) U.S. and Soviet/Russian space stations. Spending up to half a year in space with a small group of others is no longer unusual. However, plans are afoot to return to the Moon and establish a permanent settlement there and then to proceed to Mars. Big challenges are on the horizon, and their likely success is predicated on three historical series of events: first, the long series of successes in Earth-orbital flights since the launch of Sputnik on 4 October 1957; second, the six successful excursions of Apollo astronauts on the Moon; and third, the successful robotic landings to date on Mars.

In addition to the challenges that lie ahead for the big government-sponsored Moon and Mars projects of the future, the challenging era of civilian space tourism is about to begin. Five persons, beginning with Dennis Tito in 2001, have purchased flights to the International Space Station on board Russian Soyuz spaceships and have had short stays of about a week in space. On 4 October 2004, Burt Rutan and his team at Scaled Composites, LLC, won the $10 million Ansari X Prize by successfully flying a privately developed and funded suborbital rocket ship capable of reaching space. The British company Virgin Galactic is now planning to fly commercial passengers to the edge of space in a larger suborbital spaceship being developed by the X Prize–winning Scaled Composites in Mojave, California. These will be short rides in space, lasting only minutes, but it is clear that the era of space tourism is at hand. Travel and touring form a powerful human motive, an observation corroborated by the fact that tourism is the world's largest industry. Spaceflight involving humans used to be exclusively the domain of the massive governmental programs in the space race between the United States and the Soviet Union. Now, however, other countries have smaller but significant space programs with human spacefarers. Robert Bigelow's private aerospace company in Las Vegas, Nevada, now has flying in Earth orbit a pair of proof-of-concept scale models of a proposed generic habitat that could become a space hotel or a private space factory or laboratory.

Both the Soviet/Russian and U.S. space programs have demonstrated that humans—men and women of different national and ethnic groups—can live and work together in Earth-orbiting habitats for modest periods that have quite precise beginning and ending times. But all of these successful experiences have taken place in the quasi-military social structures of the astronaut and cosmonaut space programs. In addition, both of these programs managed the expected problems of human physical and social frailty by very stringent selection procedures and lengthy training regimens. None of this is compatible with space tourism. Space tourists will compose a much wider spectrum of spacefarers who will have to be accommodated and whose training periods will have to be dramatically shorter. In addition, space tourists will not be government employees and heroes of national renown. Rather, they will be purchasing their flights to space and will expect services commensurate with the cost of the tour.

I am proposing that, as we approach the end of the first half century of spaceflight, the accelerating maturation of the space program has brought us to a historical threshold. To move successfully beyond that threshold will require a significant shift in the way engineers, designers, and flight managers approach their tasks. The time is past when getting to space was a dangerous experiment; technology was in an early stage of development, and it was sufficient (and perhaps necessary) to design minimal spacecraft that required highly selected and arduously trained test pilots in top physical condition to tolerate them as environments.

At this point in the history of spaceflight, we face two big questions: 1) even though the cosmonaut and astronaut social systems have adapted to the demands placed on them so far, will they be able to cope with the much greater challenges that lie ahead, and 2) is there any chance that space tourism, with a much more fluid social structure and a vastly broader spectrum of participants than in the current space program, will work at all? This paper deals primarily with the second question, but some of the things that make civilian spaceflight possible will apply to facilitating astronaut and cosmonaut success with their new challenges of establishing a permanent Moon base and then going on to Mars.

Social scientists have been studying the behavior of humans in what have come to be called "extreme environments" even before the space program began. Extreme environments are those that are characterized by such features as isolation, risk, confinement, crowding, restricted privacy, and the inability to leave at will. Illustrations of such environments are submarines, Arctic and Antarctic research stations, undersea habitats, and high-altitude research stations. Once there was interest in how people might perform in tiny, crowded spacecraft, these studies of other extreme environments were looked at as analogs for spaceflight since all of the environments had salient social stressors in common. It seemed reasonable to assume that it did not much matter if one was deep in the ocean or up in space if one was confined in a metal canister with other people. A good summary of this work can be found in A. A. Harrison's book, *Spacefaring: The Human Dimension*.[1] Suffice it to say here that these were good analogs for what spaceflight would turn out to be like. The design of the interiors of spacecraft and the management of daily life on board benefited much from the information gleaned from experiences with other extreme environments. These experiences contributed to the development of selection criteria and training regimens in the early days of the space program. When reviewed by social scientists, these studies of real-life situations generated hypotheses and theories that were then subjected to rigorous tests in laboratories and added to the general fund of knowledge in social psychology. An example might be Freedman's density intensity hypothesis.[2] The density intensity hypothesis stated that whatever emotion is extant when crowding occurs will be enhanced by the crowding. Crowding per se is not necessarily aversive. This was a nonintuitive but valuable finding. This phenomenon can be witnessed at most happy social gatherings. A group of people may have a whole house at their disposal, but one would seldom find them uniformly distributed about the premises. It is much more likely that they will be gathered together in two or three locations, happily interacting in close proximity. The reverse of this is also true, as can be seen in mob behavior, where the crowding amplifies the anger of the members. The important point for spacecraft design was that happy people in a cramped spacecraft would not have their good moods diminished by the crowding.

Just as the study of behavior in other extreme environments turned out to be valuable for designing spacecraft interiors and predicting behavioral patterns in eventual spaceflight, so too spaceflight simulators can be used to experiment with different aspects of spacecraft interiors such as color, lighting, noise, and temperature in order to determine optimal designs before committing to a final configuration. In fact, this method began with simple cardboard and wood mock-ups (primitive types of simulators); then, beginning in the 1950s with the advent of simulators, it became commonplace in the aerospace industry and at NASA. More importantly however, spaceflight simulators can also be used to explore social behavior in extreme

---

1. Albert A. Harrison, *Spacefaring, The Human Dimension* (Berkeley: University of California Press, 2001).
2. J. L. Freedman, *Crowding and Behavior* (San Francisco, CA: W. H. Freeman, 1975).

environments. Experiments can be conducted with procedures designed to facilitate the prosocial behavior of the occupants. In the early days of the space program, when anecdotal studies of life in extreme environments such as submarines were all we had, these studies proved valuable and served us well. But spaceflight simulators can be used to create situations more specific to spaceflight and do so in a laboratory setting in which extraneous variables can be much better controlled.

Of course, spaceflight simulators on Earth cannot simulate weightlessness. That is unfortunate because the higher the fidelity of the simulator, the better the transfer to real-world situations. We have seen in aviation that extremely high-fidelity flight simulators can be so effective for training that airline pilots transitioning to a new airplane can take all of their training in a simulator and then go out and fly the airplane successfully the first time. However, the inability to simulate weightlessness by no means destroys the usefulness of spaceflight simulators as research tools for the study of human behavior. NASA currently uses them effectively to train astronauts for life on the International Space Station. Not every variable involved in an extreme environment has to be present in a simulation for an analog to be able to generalize usefully from one setting to another. If that were the case, the early studies from real-world extreme environments would have been useless to NASA. But they were not.

Now, as we enter this new era of spaceflight, we need to use simulators to improve our understanding of the variables involved in successfully coping with the new challenges that will confront us. How such simulators can be used as research tools as opposed to training devices is not intuitively obvious. Since there have been few such studies involving civilian participants, the general public knows little of what goes on in such a study. Therefore, I will describe a study conducted in my laboratory that will demonstrate how simulator studies can address both applied and theoretical research questions. This study is particularly relevant to this paper because it has given us important information about managing negative interpersonal interactions in a setting simulating those that would be found in future civilian spaceflight.

This study was conducted for McDonnell Douglas Aerospace in Huntington Beach, California, in the spring of 1996. The final report was submitted to the company and not published in the academic press. Here is an overview of the study conducted in my laboratory that shows how we might change the course of spaceflight design for the next phase in the history of life off planet Earth.

THE AEROSPACE PSYCHOLOGY LABORATORY SIMULATOR EXPERIMENT

I will begin by setting the stage for what took place. McDonnell Douglas Aerospace (now Boeing Space Systems) in Huntington Beach, California, was in the process of developing a new, single-stage-to-orbit rocket to replace the Space Shuttles. This vehicle would take off vertically the way the Shuttles do, but instead of gliding in for a landing, it would land vertically using the thrust of its engines the way the Moon landers did in the Apollo program. The rocket, which was to be called the *Delta Clipper*, was first conceived of as a cargo vehicle. Soon, engineers began thinking about having both a cargo bay and, interchangeable with it, a passenger compartment. The passenger compartment was to accommodate six passengers and a crew of two for a two-day orbital flight. Former astronaut Charles "Pete" Conrad was then a vice president of McDonnell Douglas Aerospace and a key player in the development of the *Delta Clipper*. At the time, all of the McDonnell Douglas designers were fully occupied with work under a NASA contract on the design of what would eventually become the International Space Station. Dr. William Gaubatz, who headed the *Delta Clipper* program, had hired one of my graduate students as part of the team developing the vehicle. She gave Dr. Gaubatz and Pete Conrad a copy of my book, *Human Factors in the Design of Spacecraft*.[3] After reading the book and engaging me in several interviews, they selected my laboratory to design the passenger compartment for the *Delta Clipper*.

Pete Conrad had determined that a spaceship in orbit about Earth at the inclination then being commonly flown by the Shuttles would be able to see most of the parts of Earth that the passengers would want to see in daylight if the spacecraft orbited for two consecutive days (remember that when orbiting Earth, one is in darkness half the time).

Once the passenger compartment design was satisfactorily completed, there was considerable excitement among the McDonnell Douglas engineers about the idea of taking civilian passengers to space (no one spoke words such as "space tourism" yet at that time). The designers were excited about such ideas as not putting full fuel on board the vehicle for orbital flight but keeping it lighter, adding more passengers,

---

3. H. A. Wichman, *Human Factors in the Design of Spacecraft* (Stony Brook: State University of New York [SUNY] Research Foundation, 1992).

Managing Negative Interactions in Space Crews: The Role of Simulator Research

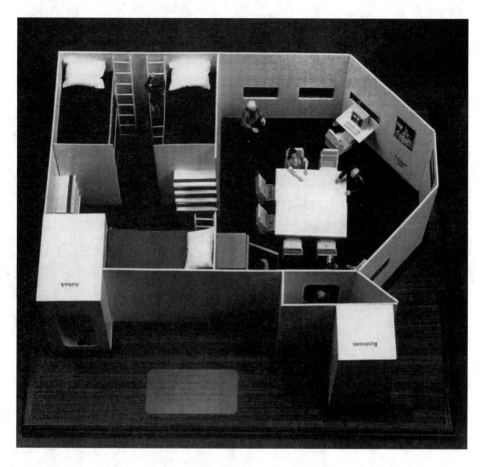

Figure 1. A scale model of the spacecraft simulator used in the study conducted in the Aerospace Psychology Laboratory at Claremont McKenna College.

remaining suborbital, and flying from Los Angeles to Tokyo in 40 minutes or Los Angeles to Paris in 38 minutes. However, when the euphoria of the daydreaming was over, the *Delta Clipper* team was left with the question, can you really take a group of unselected, relatively untrained civilians; coop them up in a cramped spacecraft for two days of orbital spaceflight; and expect them to have a good time?

A new academic year was about to begin, so the designers called a meeting and asked my lab to address this question in a simulated spaceflight.

We accepted the challenge and built a spaceflight simulator in our laboratory that had the same volume per person as the one designed for the *Delta Clipper* (see figures 1 and 2).

Psychology of Space Exploration

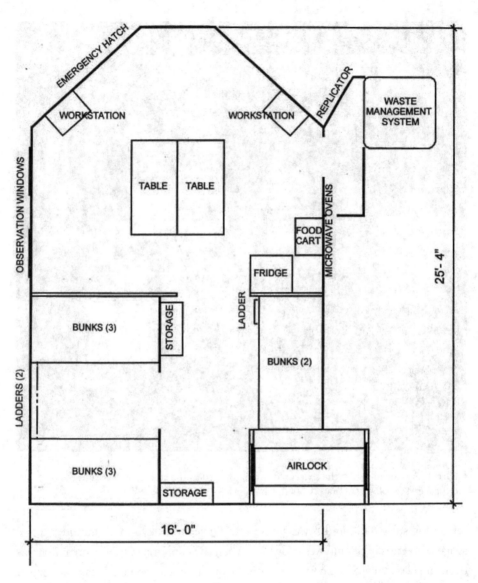

Figure 2. A diagram of the interior of the spaceflight simulator pictured in figure 1.

The *Delta Clipper* team wanted to know whether people such as those we would select could tolerate being enclosed in a simulator for 45 hours, whether this experience could be an enjoyable space vacation adventure, and whether anything could be done prior to a flight to ensure a high quality of interpersonal interactions among the participants during the flight. Furthermore, they wanted us to measure the quality of the interactions among the participants. Answering these questions required that we conduct an experiment; we decided to conduct two simulated flights with equivalent groups. The flights would have to be essentially identical except that one group (the experimental group) would get preflight training in effective group behavior techniques, and the other group (the control group) would spend the same preflight time in a placebo treatment without group training.

The two groups would not know whether they were the experimental or control group. The groups would have to be observed during the simulation. They did know that they would be observed by cameras and microphones placed so as to survey the entire interior except in the airlock and the toilet. (Participants changed clothes and "bathed" in the airlock with moist towels that were warmed in a microwave oven.) The participants could also be observed through a one-way window that appeared to be a mirror on their side. Participants soon were oblivious to being observed, as was often demonstrated when an observer on the outside would be startled by a participant suddenly using the one-way window as the mirror it appeared to be on the inside.

In an effort to recruit participants who would approximate the kinds of people who might book a spaceflight, we contacted a travel company that booked adventure travel tours such as to Antarctica and got from them the demographics of the people who book such tours. We then advertised in a local paper for volunteers to act as participants in a simulated space "vacation." Those applying would have to commit to participating for 48 hours, from 5 p.m. on a Friday evening until 5 p.m. on the following Sunday evening. Six passengers were selected for each of the two groups: they ranged in age from 34 to 72, half of them were men and half were women, and each group had one married couple. In addition, each group had its own two-member crew, a white male and a black female. We knew of no spaceflight simulation study that involved such diversity of age, gender, and ethnicity involving civilians resembling those who might one day be involved in space travel. Participants wore their own light sport clothing and soft slippers or warm socks because, as they were informed, in space, where people will be floating about and might bump into others or delicate equipment, shoes would not be worn. The crewmembers were mature college students who were recruited and trained ahead of time. They wore uniforms similar to NASA-type coveralls. They were unaware of the fact that there were two groups and of the variables being studied.

Observers were trained to a high degree of reliability to observe the groups at all times. The analytical system used was the Bales Interaction Analysis technique.[4] Using operationally defined criteria, the observers measured whether interpersonal interactions, both verbal and nonverbal (e.g., postures, gestures, and expressions), were positive, neutral, or negative.

During their duty shifts, the observers each monitored the behavior of two participants. An observer would monitor one participant for a 1-minute period, assign a score, and then switch to the other participant for a 1-minute period and assign that person a score. Then it was back to the first person for a minute and so on until the end of the shift.

Our spaceship simulator had the same shortcoming that all earthbound simulators have: it could not simulate weightlessness. However, that does not seem to be a critical factor. The astronaut and cosmonaut programs have a long history of using such simulators and getting results in actual spaceflight that match the behaviors observed in the simulators with a high degree of fidelity.

A simulator is, in a way, equivalent to a stage set. If it looks sufficiently like a spaceship and has the sounds and smells of a spaceship, and if the things that take place within it are those that take place in spaceflight, then the participants, so to speak, "buy into it" and experience the event as a spaceflight. Our spaceflight simulator seems to have worked very well in this respect. Loudspeakers produced sounds mimicking those in Space Shuttles and were kept at amplitudes similar to the Shuttle averages (72 decibels). For liftoff and touchdown, very loud engine exhaust vibration and sound were produced by large, hidden speakers.

Because the participants in the simulator did not float about in weightlessness as they would in orbit, we had to have bunks for them to sleep in. During the simulated liftoff and insertion into orbital flight, the participants remained strapped in their bunks. The participants reported in postflight questionnaires that they felt they really had a sense of what a spaceflight would be like—that they often forgot that this was "make believe" and that they "really were living the real experience," to quote two of the participants. They reported being thrilled during the noisy liftoff and the powered landing.

---

4. R. F. Bales, *Personality and Interpersonal Behavior* (New York: Holt, Rinehart and Winston, 1970).

Managing Negative Interactions in Space Crews: The Role of Simulator Research

The moment our experimental team knew for certain that we had been successful in creating a realistic spaceflight simulation occurred soon after the first group of participants was established in "orbit." The crewmembers were at their control stations and communicating with "Mission Control." The passengers had unstrapped from their bunks and were assembled in their seats facing forward toward the window area. Mission Control advised that they were preparing to remotely retract the radiation shield over the window and that everyone would soon have a view of Earth from space. By watching the changing postures of the participants, observers could easily see that tension was mounting during the 10-second countdown. Suddenly, a view of Earth taken from one of the Shuttle flights filled the window (actually a 27-inch television screen). One participant gasped and placed her hand to her mouth while staring at the scene. Another whispered aloud in awe, "Ohhhh myyy god." Another, holding her hands to her cheeks said tearfully, "Ohhh, isn't that beautiful." One of the men, gripping the sides of his chair, simply whispered, "Wow!" Another said, "Jeez, look at that!"

There is no doubt that the confinement, training, spacecraft routine, etc., were tolerated. All of the hourly questionnaires, the interaction measurements, and the postflight behavior of the participants clearly suggest to the most casual observer that the flight was not only tolerated but thoroughly enjoyed by the participants. It lived up to their preflight hopes, according to postflight questionnaire reports and their comments to us. Exit from the simulator was delayed in both flights because the participants took the unplanned-for time to trade telephone numbers and addresses before leaving. People who came to the experiment as strangers left as friends. Participants' moods during the simulated flights remained positive, and the number of negative interactions in both groups was small. In the year following the study, my lab received so many telephone calls from participants in both groups requesting a reunion that we felt compelled to go back to McDonnell Douglas and request that they sponsor such an event. They did, and it was a very well-attended, robust party.

One might then wonder if perhaps the social situation produced happiness but the space aspect of it was not important. What we learned is that the participants returned home and presented themselves in their respective social groups as having had a virtual trip to space of such realism that for all practical purposes, it might as well have been real. They were now, in their respective social settings, authorities on space travel. For months after the experiment, the lab kept receiving calls from participants requesting answers to all sorts of space questions. It seemed that now that they were perceived by others as authorities on space, people called to ask space-related questions of them. When they could not answer them, they turned to us for the answers. What is important here is that this postsimulation experience gave us the opportunity to see how this simulation had changed the participants' lives in a positive, space-related way.

During the first of the two simulations, we had programmed in an hour during the early part of the "orbiting" for the participants to talk with "Mission Control." This was a question-and-answer period. We hoped they might find it instructive and even entertaining. As it turned out, the early part of their flight had generated many questions that they were eager to have answered or have an opportunity to discuss. They so enjoyed that hour that they pleaded to have another such hour added for later in the flight. Fortunately, this request was made on the first flight, because the experimental design required that the program schedules be the same in both flights. We did add the extra hour to both flights. In the postflight questionnaires, the participants of both flights indicated that the discussions with Mission Control while flying were the favorite parts of the trip. Very clearly, the participants enjoyed the spaceflight aspects of the simulation very much. All of the subsystems of the simulator worked as planned. No extraneous variables intruded, such as outside noises. From the standpoint of the equipment, the experiment was uneventful.

Just before entering the simulator, the experimental group received a 2-hour-long program designed to enhance interpersonal prosocial behavior. It was designed much like the type of program corporations provide for their executives in order to develop team building and enhance effective workplace interactions. The program was divided into two main sections, one emphasizing effective interpersonal behavior and the other dealing with understanding and resolving conflicts. In essence, the first part taught participants how to be sensitive to one another and get along, and the second part taught them how to recover if a falling-out did occur. Each participant in the experimental group received a bound 14-page pamphlet of notes about the course material. Much of the formal presentation was lecture/discussion in style, but about one-fifth involved group activities as well.

Just before entering the spacecraft, the control group was given a presentation that lasted the same amount of time but had nothing to do with effective group behavior.

Table 1. The total number of interpersonal interactions and their emotional nature (positive, neutral, or negative) for participants in the experimental and control groups.

Type of Communication	Experimental Group	Control Group	Percent Difference
Positive Communication	354	282	20
Negative Communication	8	19	58
Neutral Communication	2,120	2,370	11
Total Communication	2,482	2,671	7.1

This pair of simulation studies provided much useful information. First of all, as has already been indicated, the participants not only tolerated their confinement very well, but really did enjoy it as a simulated space adventure. The observation data showed that the emotional tone of most of the interpersonal interactions was neutral and that there were relatively few negative interactions in either group, but there were significantly more negative interactions in the control group than in the experimental group that had received the preflight training in effective group behavior. Since the total number of interactions differed only slightly between the two groups, that result also meant that the experimental group had more positive interactions than the control group that received the placebo training. Table 1 summarizes the interaction data from the study. There were 2,482 total communications in the experimental group and 2,671 in the control group, a difference of only 7.1 percent.

In order to convey the basic meaning of data such as those above in a simplified manner that would also allow an easy comparison of the difference between two groups, we developed a metric called the index of amicability. This index compared the number of positive and negative interactions in the form of a ratio.[5] An amicability index of 1 means there are equal numbers of positive and negative interactions. An index smaller than 1 (e.g., 0.75) indicates that there are more negative

---

5. The index of amicability is the ratio of positive to negative interactions for a given group, or, $AI = P/N$, where AI is the amicability index, P is the number of emotionally positive interactions, and N is the number of emotionally negative interactions.

than positive interactions. An index greater than 1 (e.g., 25.0) shows that there are more positive than negative interactions. In our simulations, the experimental group index of amicability was 44.3. The control group had an index of amicability of 14.8. Thus, using the difference in index of amicability between the two groups as a measure of the efficacy of the preflight training, we find a very large improvement in social functioning of 299 percent from a small investment of 2 hours in a training program.

It is important to remember that the total number of negative interactions was low in both groups and that both groups enjoyed the experience very much. There were no nasty incidents in either group. However, the group with the preflight training had an index of amicability nearly three times greater than that of the placebo group.

It is necessary to report publicly the above civilian spaceflight simulation study at this time in order to employ its results in discussing the management of negative interpersonal interactions. As the impetus for space tourism ramps up, more such studies will be forthcoming that will attempt to replicate these findings. They will also greatly broaden the range of variables explored. We now have useful information, especially about simulated civilian spaceflight, to use in discussing all interpersonal interaction—positive, neutral, and negative. One of the purposes of this paper is to excite other scientists to conduct such research. Bales and others have given us the tools to be able to perform interpersonal interaction studies. This report of the study conducted in my laboratory demonstrates that such studies can be accomplished and produce valuable results.

Shortly before we conducted the simulation study in my laboratory, Sandal, Vaernes, and Ursin reported a simulation study of long-duration spaceflights (30 and 60 days) that they had conducted for the European Space Agency (ESA).[6] This group of researchers used decompression chambers at a naval base as spaceflight simulators. This study, too, used the Bales Interaction Analysis technique. We patterned our design after theirs so that our data could be compared. Prior anecdotal evidence from extreme environments (e.g., wintering over in Antarctica) suggested that negative interpersonal interactions among persons in the environment tended

---

6. G. Sandal, R. Vaernes, and H. Ursin, "Interpersonal Relations During Simulated Space Missions," *Aviation, Space, and Environmental Medicine* 66, no. 7 (July 1995): 17–24.

to peak at the midpoint and shortly before the end of a stay.[7] If that were the case it, would be important information for flight managers and participants to know in understanding and managing spaceflight events.

Here was an illustration of using a simulator as a laboratory to subject anecdotal analog information to experimental testing to establish more reliably the anecdotal information. The researchers in the ESA study in fact found that the anecdotal information was true and held, independent of the duration (30 or 60 days).

The *Delta Clipper* team was interested in much shorter timespans than those with which the long-duration studies had dealt. They wanted to know if this same phenomenon held for short periods, too, such as the two-day simulation we were conducting for them. We designed our study to test whether the negative interactions in our groups peaked at the midpoints and just before the ends of the simulated flights. We found that our short-duration experimental study corroborated the findings of both the long-duration experimental studies and the anecdotal studies.

We then had two experimental studies that confirmed the anecdotal findings that negative interpersonal interactions peak at the middle and near the ends of group activities in which the durations of the events are known to the participants. What is more, the finding was independent of the durations involved. It held for short periods, as found in the study reported here, and longer periods, as reported in the ESA study. This information was useful beyond the realm of spaceflight and probably generalizes to all social epochs such as family vacations and school semesters, even if the participants are not confined.

Both of these studies reported that the great preponderance of interpersonal interactions was neutral and that positive interactions were much more frequent than negative ones. Both studies reported that people got along quite well in simulations of differing durations, with more formal and less formal social structure, and in private as well as government settings. The results of the study in my laboratory should be good news for the neophyte space tourism industry. It showed not only that relatively unselected, minimally trained civilians can tolerate the extreme environment of a spaceflight simulator, but also that they find the experience pro-

---

7. A. A. Harrison and Y. A. Clearwater, eds., *From Antarctica to Outer Space: Life in Isolation and Confinement* (New York: Springer-Verlag, 1991).

foundly pleasing. In fact, these results hint that one could build a viable business out of just selling simulated spaceflights.

Another encouraging finding of the study reported here is the powerful effect of preflight group dynamics training on reducing negative interactions and increasing those that are positive. We are reminded by Freedman's Density Intensity Hypothesis that crowding is not necessarily an aversive stimulus, but that it does tend to amplify whatever emotion is extant in a group.

Research that answers questions invariably raises new ones, and that is true of this research. Here is a sample of some questions raised by this research:
1. For how long does the effect of preflight training last? In this project, it only had to persist for 45 hours.
2. Does the effect of preflight training end abruptly, or does it taper off?
3. Does a small increment of booster training return preflight training to its original effectiveness?
4. Which of the variables involved in the preflight training are responsible for the effect it produced?
5. Are there other variables that could be added to the preflight training syllabus that would increase its positive effect or duration or both?

A CALL FOR A DIFFERENT TYPE OF THINKING

As the history of spaceflight unfolds, I contend that now we are at a transition point between the exploratory and settlement stages of spaceflight that is similar to the opening of the American West in the United States. The early exploration of the West was conducted by a relatively few brave and hardy explorer sorts with an emphasis on daring and pushing back frontiers. There was much ambiguity about the challenges and dangers that lay in uncharted territory. These beginning forays into the unknown were followed by the incursion of hardy trappers, hunters, miners, and various tradesmen. Settlers soon followed, and eventually tourists did as well. In parallel with the western movement of people, technology was improving to facilitate the western expansion—transportation evolved from stage coaches and Conestoga wagons to steamboats and trains.

Managing Negative Interactions in Space Crews: The Role of Simulator Research

So, too, the early stages of spaceflight were conducted by heroic persons such as Yuri Gagarin and John Glenn. Space stations were eventually established in the frontier, and people learned to adapt to life in space. Now we are planning to return to the Moon and establish a permanent settlement there. The newest frontier dream, the planet Mars, is several orders of magnitude distant from the Moon. We have an International Space Station with a short but significant history of long-duration multinational crews. It has even been visited several times by tourists. In the beginning, the attitude about equipment design was simple: get there, survive, and get back in one piece. So too with astronaut selection: prove that you have "the right stuff"—which translated to "be a high-performance test pilot."

Our technology is much more sophisticated now than it was in the early days of the space program. Now we are designing vastly larger and more complex space stations, Moon colonies, long-duration spaceships for the journey to Mars, and space hotels. The people we will be sending to these sites will be scientists, technicians, service persons, and tourists. I believe that the shift in emphasis implied by these changes requires a shift in the way we think about space equipment and the personnel who will use that equipment.

THINKING ABOUT EQUIPMENT

It is no longer enough to design to survive.
The time has come to design to thrive.

In the beginning of the space program, engineers were not eager to have people on board space vehicles. The design spirit seemed to be something like, "We are confident this will work, so let's use it. If it is uncomfortable or it is difficult to operate, then find astronauts who can tolerate it and who can be trained to make it work." We now need to shift design thinking to a human factors and ergonomics point of view. This kind of shift in emphasis means designing the apparatus to match the capabilities and aspirations of those who will use it. For example, space vehicles are currently very noisy. The noise is due to the fact that warm air does not rise in weightless environments. Without convection currents, any air that is to be moved must be moved mechanically. The large number of fans and bends in ductwork create much of the noise. Spacecraft typically have sound levels of about

72 db.[8] This is about like driving a car at 100 kilometers per hour (kph) with the windows rolled down. By comparison, a living room on Earth would be about 45 db.

This is much too noisy for comfort over long durations. Such noise levels degrade performance, communication, and satisfaction.[9] Another human factors issue is spaciousness and privacy. Once the Space Shuttles became the primary heavy-lift spacecraft for the United States, the size of their cargo bays became the limiting factor for space hardware. Thus, the U.S. Destiny module on the International Space Station is 4.3 meters in diameter and 8.5 meters long. All of the other modules are similar. Fortunately, we are on the verge of having space modules considerably larger in volume than those that existed on the Soviet space station *Mir* or those currently on the International Space Station. These are inflatable modules, such as NASA's Transhab[10] and other structures based on it, that are being privately developed by Bigelow Aerospace in Las Vegas, Nevada. Bigelow's Genesis I (launched 12 July 2006) and Genesis II (launched 28 June 2007)[11] are currently in orbit and functioning as planned. NASA canceled the Transhab program in 2000, but development work (based on NASA's efforts) continues at Bigelow Aerospace, and that is encouraging. The Genesis modules are both prototype, proof-of-concept structures in flight at the present time. Both models of Genesis were launched on Russian rockets and then inflated in space. Having a crowded cabin on a spaceship transporting people to space is no problem; after all, it is only a 100-mile trip. But living for extended periods of months with little privacy and cramped quarters, while obviously tolerable (as on the International Space Station), is not comfortable. The efforts of Bigelow

---

8. H. A. Wichman, "Designing User-Friendly Civilian Spacecraft," paper 95-604 in *Proceedings of the 6th International Space Conference of Pacific Basin Societies* 91 (December 1995), available online at *http://www.spacefuture.com/archive/designing_user_friendly_civilian_ spacecraft.shtml* (accessed 18 June 2007).

9. Paul A. Bell, Thomas C. Greene, Jeffrey D. Fisher, and Andrew Baum, *Environmental Psychology*, 5th ed. (Fort Worth, TX: Harcourt Brace, 2003).

10. K. Dismukes (curator), "Transhab Concept," *International Space Station History*, *http:// spaceflight.nasa.gov/history/station/transhab/* (accessed 3 July 2007).

11. E. Haakonstad, "Genesis II Different from Genesis I," *Out There*, *http://web.archive.org/ web/20070528014400/http://www.bigelowaerospace.com/out_there/genesis_II_difference.php* (accessed 3 July 2007).

Aerospace to move beyond the limits of past equipment design is illustrative of the shift in thinking that I am proposing as timely.

THINKING ABOUT SPACEFARERS

Effective behavior stems not from "good" people
It is called forth from "good" environments

In the early days of the space program, little was known about the effects of spaceflight on humans, physically or mentally, and the equipment was rudimentary. At that time, it made sense to experiment only with rigorously selected individuals who were exceptional physical and mental specimens. Those days are now over. We are about to enter an era of space tourism. The great message of social psychology is that the behavior we usually attribute to our character is much more determined by our environment than we ever imagined. This finding was amply demonstrated in the following three projects.

In his famous study of obedience, Stanley Milgram showed that everyday Americans could be made to behave cruelly by the making of subtle changes in an academic-like environment.[12]

In an infamous 1964 murder, a young woman named Kitty Genovese was slowly killed through the night while she pleaded for help, but no one came to her aid or even called the police. Many of the people living in her apartment building admitted hearing her but were not motivated to help. Shortly thereafter, John Darley and Bibb Latané began their classic studies of bystander intervention and clarified the social and environmental variables that call forth or inhibit bystander intervention no matter who the bystander may be.[13]

Finally, Philip Zimbardo, in his classic 1971 Stanford Prison experiment, showed how social circumstances could cause a randomly assigned group of

---

12. Stanley Milgram, "Behavioral Study of Obedience," *Journal of Abnormal and Social Psychology* 67 (1963): 371–378.
13. John Darley and Bibb Latané, *Journal of Personality and Social Psychology* 8 (1968): 377–383.

Stanford students to be so cruel to another randomly assigned group that the study had to be terminated.[14]

The three projects described are dramatic because they dealt with negative behavior. But in the study reported in this paper, with only a 2-hour training program for essentially unselected people of a wide range of ages, we were able to produce an index of amicability in one group that was nearly three times greater than that in an equivalent group. The argument here is not against any selection. Obviously passengers in an airliner want their pilot to have good vision and a healthy heart. The argument here is for a shift in emphasis toward making it possible for a broad spectrum of people to become space tourists by briefly but effectively developing in them the social skills necessary for a safe and pleasant experience in space.

The spaceflight simulator is an excellent tool, both in which to conduct the necessary social psychology research to show what needs to be done and as the vehicle for conducting the training exercises to bring that about. Interestingly, in terms of selection, the spaceflight simulator provides people with an opportunity to see if they will like such an experience; if not, they will select themselves out without jeopardizing the safety or happiness of others on a real spaceflight.

There is a subset of social psychology theory referred to as attribution theory. Much of the research in this area indicates that humans have a tendency to attribute people's behavior to their character. This is known as the fundamental attribution error. It is the tendency to over-attribute the motivation for a person's behavior to that person's character and underestimate the effect of situational factors. When we emphasize selecting the right "type" of person for spaceflight instead of creating the right "type" of social and environmental factors, we are committing the fundamental attribution error. We have seen in the three social psychology research projects cited above how "good" people could be made to do "bad" things by simple manipulation of situational circumstances. We have also seen in my Aerospace Psychology Laboratory study presented here that people in one group similar to people in another group could have the negative behaviors they would be expected to produce dramatically reduced (58 percent; see table 1) by a small amount of focused training. Emphasizing selection will diminish the number of prospective spacefarers and inhibit the growth of space tourism. However, emphasizing environmental design and training instead

---

14. David G. Meyers, *Social Psychology*, 9th ed. (New York: McGraw-Hill, 2006).

will have the effect of broadening the spectrum of potential spacefarers, facilitating the growth of space commercialization, and, finally, increasing the satisfaction everyone experiences from spaceflights. The primary tool available for fulfilling this shift in emphasis is the spaceflight simulator.

As this paper is being written, the European Space Agency has just issued a worldwide invitation for volunteers to participate in a 520-day simulated Mars mission. Let us hope that this is only the beginning of a long series of studies that will reflect a fundamental change in the way the aerospace industry thinks about the behavior of people in space vehicles and habitats.

Chapter 6
Gender Composition and Crew Cohesion During Long-Duration Space Missions

Jason P. Kring
　Department of Human Factors and Systems
　Embry-Riddle Aeronautical University

Megan A. Kaminski
　Program in Human Factors and Applied Cognition
　George Mason University

ABSTRACT

A major factor in the success of future long-duration space missions is the psychosocial functioning of the crew. An individual's psychological health and well-being has a major impact on how well he or she adapts to the demands of isolation, confinement, and workload associated with complex missions. Although each crewmember possesses a unique combination of knowledge, skills, and abilities that influence their capacity to adapt, in this chapter we argue that mission success also relies on how well an individual functions in the larger social context of the mission. More specifically, interactions between crewmembers, as well as between the crew and ground personnel, play a significant role in the crew's overall performance. Although many variables affect crew interactions, such as opportunities for personal space and privacy afforded by the spacecraft's architecture, we contend that the most prominent factor is the crew's composition. Beyond the size of the crew, the mixture of cultural and ethnic backgrounds, and the blend of professional expertise, the most salient crew composition variable is gender.

Since even before Valentina Tereshkova's flight in 1963, women have played an integral role in the history of human spaceflight. As of April 2010, for instance, 53 different women have flown in space, many as part of mixed-gendered crews aboard Russian space stations or the International Space Station (ISS). The April 2010 flight of Space Shuttle *Discovery* to the ISS set a record for the most women in space at one time as three female crewmembers aboard *Discovery*—Dorothy Metcalf-Lindenburger, Stephanie Wilson, and Naoko Yamazaki—joined Station resident Tracy Caldwell Dyson in orbit. As the number of mixed-gender crews will likely increase in the future, including those taking voyages to near-Earth asteroids and then to Mars, it is prudent to ask if there are any potential limitations to men and women working together for extended periods of time.

This chapter reviews findings from mixed-gender crews in spaceflight as well as relevant analogs like aviation, Antarctic research bases, and other complex environments to highlight how gender composition moderates crew interactions and performance. To explore this relationship, we focus specifically on the variable of cohesion, or the degree to which crewmembers are committed to each other and to the crew's shared task, and offer recommendations for the optimal gender composition for future space missions in terms of this important crew variable.

INTRODUCTION

In 2004, the Bush administration's Vision for Space Exploration refocused the U.S. human spaceflight program on returning people to the Moon by 2020 and then sending a crew to Mars. The Obama administration's plan, announced in April 2010, focuses on asteroid rendezvous missions as stepping-stones for a Mars flight. Regardless of the specifics of the plan, sending humans beyond low-Earth orbit is ambitious on many fronts and will require the development of a host of new technologies, from improved launch and propulsion systems to a completely new crew vehicle. Scientists and engineers must also work diligently to design systems and mission activities to protect against physiological risks associated with long-duration spaceflight (LDSF), including radiation exposure, bone degradation, and muscle loss. However, beyond the technical and physiological challenges, a major obstacle to LDSF is the psychosocial environment during the mission. In conjunction with individual responses to isolation and confinement, researchers contend that problems associated with crew interactions may be a significant limiting factor for extended space missions.[1] Jack

---

1. Nick Kanas and Dietrich Manzey, *Space Psychology and Psychiatry* (El Segundo, CA: Microcosm Press and Kluwer Academic Publishers, 2003).

Stuster emphasizes this point in his review of literature from spaceflight and similar domains like polar bases and stations. Based on the writings of behavioral scientists and accounts from explorers and Antarctic personnel, he concludes that "the smooth functioning of the group contributes greatly to mission success and can be essential to survival under emergency conditions."[2]

The factors influencing something as dynamic and complex as human interaction are, of course, numerous. For even the simplest one-on-one conversation, the personalities of the individuals, their motivations, their organizational roles (e.g., leader versus follower), and the context of the conversation affect each person's perception and interpretation of the interaction. Nonverbal cues, such as body posture, and paralinguistic cues, like the tone of voice, also shape the interaction, altering the degree to which the conversation is deemed pleasant, threatening, or productive. However, in the context of a long-duration space mission with three or more crewmembers, the complexity of human interaction increases significantly. For one thing, interactions occur in the context of a high-workload and high-stress environment. The crew is under tremendous pressure to perform tasks correctly and according to strict timelines with little room for error, creating a setting ripe for tension. Furthermore, beyond their own internal interactions, the crew must also routinely communicate with numerous groups on the ground. These can range from flight-related personnel (e.g., flight controllers, engineers, medical staff), to family members, to even representatives of the media and governments around the world.

Nevertheless, the most prominent factor affecting crew interactions is the composition of the crew itself. Findings from the behavioral and social sciences, spaceflight, and similar settings indicate that the number of people on a team or crew and their individual characteristics are influential to the team's interactions and success. Evidence from spaceflight and analogous settings like Antarctica and submarines indicate, for instance, that the size of the crew has a major impact on crew interactions. Harrison concludes that larger crews possess several advantages over smaller crews, such as a greater range of skills and abilities, as well as providing more oppor-

---

2. Jack Stuster, *Bold Endeavors: Lessons from Polar and Space Exploration* (Annapolis, MD: Naval Institute Press, 1996), p. 165.

tunities to form friendships and create a more interesting social experience.[3] Larger crews also appear to get along better, exhibit less hostility, be more stable, and make better and more efficient decisions, particularly if they are odd-numbered, because in the event of a tie, one crewmember can cast the deciding vote.[4]

In addition to the size of the crew, crew composition also refers to the characteristics of the individual members. Each crewmember brings his or her own unique qualities to the crew based on his or her experiences; attitudes; personality; motivation; and combination of knowledge, skills, and abilities. For example, differences related to national culture and ethnic background are important crew composition considerations. Although not a significant problem, there have been cases in which cultural diversity led to difficulties in crew interactions. For instance, U.S. astronauts cited cultural factors related to personal hygiene and housekeeping practices as partially responsible for incidents of miscommunication and interpersonal conflict before, during, and after international Space Shuttle missions based on responses to a survey conducted by Patricia Santy and colleagues.[5] Nine respondents—astronauts from flights between 1981 and 1990—reported over 40 incidents of misunderstanding and interpersonal friction related to culture, with at least five rated as having a high impact on the mission.

This chapter focuses on the most salient crew composition characteristic that influences crew interactions: the gender of the individual crewmembers. For this discussion, it is worthwhile to reiterate the distinction between the terms "sex" and "gender." As Stephen Davis and Joseph Palladino note, "sex" is a biological classification of male or female, whereas "gender" refers to the cultural and social expectations about what is masculine or feminine.[6] In the context of interpersonal relations, these gender-based expectations significantly influence how a man or woman interacts with others, from styles of verbal and nonverbal communication to the expression and interpretation of emotions. Although gender differences have

---

3. Albert A. Harrison, *Spacefaring: The Human Dimension* (Berkeley: University of California Press, 2001), p. 138.
4. Kanas and Manzey, *Space Psychology and Psychiatry*, pp. 87–88.
5. Patricia Santy, Albert Holland, L. Looper, and Regina Marcondes-North, "Multicultural Factors in the Space Environment: Results of an International Shuttle Crew Debrief," *Aviation, Space, and Environmental Medicine* 64 (1993): 196–200.
6. Stephen F. Davis and Joseph J. Palladino, *Psychology* (Upper Saddle River, NJ: Pearson, 2007).

the potential to affect a number of variables related to how crews interact and perform, we will address how gender, specifically in the mixture of men and women, influences the cohesion of the crew. Cohesion, defined here as the degree to which individuals in a crew or on a team are committed to each other (interpersonal cohesion) and to the goals of the team's task (task cohesion), has received considerable empirical attention. Although there is some disagreement over the specific relationship between cohesion and performance, the general consensus is that teams possessing higher levels of cohesion function more effectively and exhibit better performance than low-cohesion teams.

This chapter reviews findings from mixed-gender crews in spaceflight as well as relevant analogs like aviation, Antarctic research bases, and other complex environments to highlight how gender composition moderates crew cohesion and, ultimately, performance. The discussion also addresses challenges associated with interpersonal relationships during LDSF and whether guidelines are needed to limit or prohibit romantic relationships. We first describe what is known about mixed-gender teams in space and similar settings and then later turn to the issue of team cohesion.

WOMEN AND MEN IN SPACE AND ANALOGOUS SETTINGS

Even before Valentina Tereshkova's 1963 flight aboard Vostok 6, women have played an integral role in the history of human spaceflight. According to NASA's History Division, 53 different women have flown in space as of April 2010, including Soviet/Russian cosmonauts, American astronauts, and citizens of other countries, with 47 of these women flying with NASA.[7] Even before the U.S. program got off the ground, the Woman in Space program proved that women could endure the rigors of astronaut selection. Thirteen female pilots passed the same physical examinations at the Lovelace Clinic in New Mexico used to screen and select male pilots for the Mercury program in the late 1950s and early 1960s.[8] In space, it

---

7. NASA History Division, "Women in Space," NASA, available at *http://history.nasa.gov/women.html* (accessed 7 June 2010).

8. NASA History Division, "Lovelace's Woman in Space Program," NASA, available at *http://history.nasa.gov/flats.html* (accessed 7 June 2010).

was not until the early 1980s that the number of female astronauts and cosmonauts began to rise. In fact, nearly 20 years passed between Tereshkova's record-making flight and Svetlana Savitskaya's 1982 mission aboard a Russian Soyuz, which was followed closely by the flight of the first U.S. woman in space, Sally Ride, in 1983. Today, it is not uncommon for women to conduct extravehicular activities; pilot or command the Space Shuttle, as Eileen Collins has done twice; or command the ISS, as in the case of Peggy Whitson.

With the increase in female astronauts and cosmonauts, the number of mixed-gender crews has also risen. As of 2009, seven crews of men and women had worked together during long-duration space station missions aboard the USSR's *Salyut* station, the Soviet/Russian station *Mir*, or the ISS. As the number of mixed-gender crews will likely increase in the future, including for voyages to asteroids and then to Mars, it is prudent to ask if there are any potential limitations to men and women working together for extended periods of time.

On the surface, this mixture would seem ideal as each gender offers unique and complimentary skills and abilities to a mission. The behavioral literature is replete with studies of gender differences in cognition, sensation and perception, and team performance in business settings; however, few studies have focused specifically on gender differences in the context of extended space missions. One question is how the mixture of men and women in space crews affects overall crew performance. In the general team literature, for example, findings suggest that men and women do work in slightly different ways that may influence team performance such as leadership style and reactions to stress.[9] Research also suggests that the unique contributions from each gender often improve team performance in settings such as health care, manufacturing, and extreme environments, thereby supporting the use of mixed-gender teams.[10]

For instance, in some contexts, all-male teams make less accurate and more overly aggressive decisions than mixed-gender teams.[11] However, these and other

---

9. A. H. Eagly, M. C. Johannesen-Schmidt, and M. L. van Engen, "Transformational, Transactional, and Laissez-faire Leadership Styles: A Meta-analysis Comparing Men and Women," *Psychological Bulletin* 129, no. 4 (2003): 569–591.

10. S. L. Bishop, "Evaluating Teams in Extreme Environments: From Issues to Answers," *Aviation, Space, and Environmental Medicine* 75, no. 7, sect. II (2004): C14–C21.

11. J. A. LePine, J. R. Hollenbeck, D. R. Ilgen, J. A. Colquitt, and A. Ellis, "Gender Composition, Situational Strength and Team Decision-making Accuracy: A Criterion

results are qualified by the fact that gender differences in team performance are often moderated by other factors such as the type of task and the personality composition of the individual team members. With regard to personality in team sports, for example, male and female athletes exhibit different personality profiles and attitudes toward recreational activities.[12] Furthermore, gender heterogeneity may influence the development of team factors that contribute to successful team performance like cohesion and trust. As noted above, cohesion is a team's commitment to a shared task and attraction between team members, whereas trust refers to attitudes held by team members regarding their emotional closeness with, and the reliability of, other members of the team. How men and women respond to stress, for instance, can influence both cohesion and trust, particularly at the interpersonal or emotional level. N. S. Endler notes that men tend to cope with stress using "fight or flight" strategies, whereas women employ a "tend or befriend" approach.[13] This latter strategy may therefore evoke more emotional closeness among crewmembers.

In spaceflight, mixed-gender crews have flown successfully since the 1980s; however, a majority of these missions were short-duration flights of one to two weeks. For example, although no performance issues were attributed to Svetlana Savitskaya's gender when she visited the Soviet *Salyut 7* station for eight days in 1982, cosmonaut Valentin Lebedev's account of the visit suggests that gender stereotyping did occur. After presenting her with a floral print apron upon her arrival, he declared, "Look, Sveta, even though you are a pilot and a cosmonaut, you are still a woman first. Would you please do us the honor of being our hostess tonight?"[14] For longer-duration missions lasting between five and seven months, anecdotal reports from two mixed-gender missions aboard the ISS indicate that the crew also got along and functioned effectively.[15]

Decomposition Approach," *Organizational Behavior and Human Decision Processes* 88, no. 1 (2002): 445–475.

12. Bruce D. Kirkcaldy, "Personality Profiles at Various Levels of Athletic Participation," *Personality and Individual Differences* 3, no. 3 (1982): 321–326.

13. N. S. Endler, "The Joint Effects of Person and Situation Factors on Stress in Spaceflight," *Aviation, Space, and Environmental Medicine* 75, no. 7, sect. II (2004): C25.

14. Valentin Lebedev, *Diary of a Cosmonaut: 211 Days in Space* (New York: Bantam Books, 1988), p. 191.

15. Kanas and Manzey, *Space Psychology and Psychiatry*, p. 77.

Psychology of Space Exploration

More scientific observations of mixed-gender crews are available from research in space analogs—extreme settings that possess similar environmental and social features. In her review of team literature in extreme environments, S. L. Bishop found support for using mixed-gender crews to improve long-duration performance, concluding, "The presence of both men and women appears to normalize group behavior in ways that promote individual and group functioning."[16] This result may stem from differences each gender possesses that benefit team performance in specific situations. For example, groups composed of both men and women perform well on tasks requiring diverse perspectives and interpersonal skills, and cohesion appears to increase due to women's interpersonal style and ability to involve all group members in the task.[17]

Additional support for employing mixed-gender teams comes from Stuster's findings that although the inclusion of women at U.S. Antarctic stations in the early 1980s resulted in some minor conflicts, in general, gender diversity had a positive influence on morale and productivity.[18] E. Rosnet et al. found a comparable beneficial effect of mixed-gender groups at a French polar station; however, some women reported problems related to rude behavior from their male colleagues and instances of sexual harassment.[19] Interpersonal relationships between men and women and sexual issues may also impact mixed-gender crews during extended isolation. Crewmembers of an experiment termed the Simulation of Flight of International Crew on Space Station (SFINCSS) reported increased crew tension after an incident in which a male Russian commander from one group attempted to kiss a female Canadian crewmember from another group during a New Year's Eve celebration in the enclosed space station mock-up.[20]

In summary, despite the potential for conflict and tension, evidence from spaceflight and related environments suggests that the inclusion of men and women

---

16. Bishop, "Evaluating Teams in Extreme Environments": C17.
17. Ibid., p. C17.
18. Stuster, *Bold Endeavors*, p. 178.
19. E. Rosnet, S. Jurion, G. Cazes, and C. Bachelard, "Mixed-gender Groups: Coping Strategies and Factors of Psychological Adaptation in a Polar Environment," *Aviation, Space, and Environmental Medicine* 75, no. 7, sect. II (2004): C10–C13.
20. Gro M. Sandal, "Culture and Tension During an International Space Station Simulation: Results from SFINCSS '99," *Aviation, Space, and Environmental Medicine* 75, no. 7, sect. II (2004): C44–C51.

on long-duration space missions will benefit individual and crew functioning. However, more detailed analyses are needed to identify team performance issues specifically influenced by the gender of the crewmembers. One important question is how gender heterogeneity affects the development of crew cohesion.

COHESION

In general, cohesion refers to the closeness and solidarity of a group or team of individuals. However, researchers have long debated the specifics of the construct, particularly the number of associated factors or dimensions. Early researchers used a multidimensional approach. L. Festinger, for instance, defined cohesiveness as "the resultant of all the forces acting on the members to remain in the group. These forces may depend on the attractiveness or unattractiveness of either the prestige of the group, members in the group, or the activities in which the group engages."[21] In other words, cohesion was seen to result from one or more of three sources: group prestige, interpersonal attraction, or attraction to the group's tasks. Similarly, C. W. Langfred conceptualized cohesion as the degree to which group members feel a part of the group and their desire or motivation to remain in the group.[22] In a military context, G. L. Siebold and D. R. Kelly posited that cohesion "is a unit or group state varying in the extent to which the mechanisms of social control maintain a structured pattern of positive social relationships (bonds) between unit members, individually and collectively, necessary to achieve the unit or group's purpose."[23] In contrast, some have argued that cohesion only encompasses one dimension. Cartwright and others, for example, defined cohesion simply as the degree to which group members desire to remain in the group.[24] Similarly, Kenneth Dion defined

---

21. L. Festinger, "Informal Social Communication," *Psychological Review* 57 (1950): 274.
22. C. W. Langfred, "Is Group Cohesiveness a Double-edged Sword? An Investigation of the Effects of Cohesiveness on Performance," *Small Group Research* 29 (1998): 124–143.
23. G. L. Siebold and D. R. Kelly, *Development of the Combat Platoon Cohesion Questionnaire*, ARI Technical Report 817, ADA 204917 (Alexandria, VA: U.S. Army Research Institute for the Behavioral and Social Sciences, 1988).
24. D. Cartwright, "The Nature of Group Cohesiveness," in *Group Dynamics: Research and Theory*, ed. D. Cartwright and A. Zander, 3rd ed. (New York: Harper and Row, 1968).

group cohesion as "the social glue that binds members of a group and keeps them together in the face of internal and external threats."[25]

More recently, several authors have begun differentiating between the social or interpersonal aspects of cohesion and those related to the group task. Interpersonal cohesion includes dimensions such as interpersonal attraction and the intensity and positive nature of relationships.[26] Task cohesion, in contrast, refers to the attraction or commitment to the group and task. Task-cohesive groups, according to S. J. Zaccaro, J. Gualtieri, and D. Minionis, also "care about the success of other group members because their own goal attainment is often inextricably bound to the collective achievement. They will exert strong effort on behalf of the group and their fellow members to facilitate group processes."[27] Combining these two dimensions, we can define cohesion as the combination of task cohesion, referring to the degree to which group or team members are committed to the task, and interpersonal cohesion, the degree to which individuals are attracted to each other and have positive relationships. Before addressing this construct in terms of gender differences, it is worth briefly noting what is known regarding cohesion and performance.

After decades of research, dating back to the 1950s, the relationship between cohesion and team performance continues to generate debate. Early efforts concluded that group productivity and cohesiveness were not clearly related.[28] For example, R. M. Stogdill found, in his review of 34 studies, that roughly a third of the studies showed cohesive groups to be more productive, with a third reporting that cohesive groups were less productive and the remaining third showing no difference.[29] However, none of the studies referenced by Stogdill used the same definition for group cohesion, and many made no attempt even to measure cohesiveness. Later research showed support for a positive correlation between cohesion and per-

---

25. Kenneth L. Dion, "Interpersonal and Group Processes in Long-Term Spaceflight Crews: Perspectives from Social and Organizational Psychology," *Aviation, Space, and Environmental Medicine* 75, no. 7, sect. II (2004): C39.

26. S. A. Carless and C. De Paola, "The Measurement of Cohesion in Work Teams," *Small Group Research* 31, no. 1 (2000): 71–88.

27. S. J. Zaccaro, J. Gualtieri, and D. Minionis, "Task Cohesion as a Facilitator of Team Decision Making Under Temporal Emergency," *Military Psychology* 7, no. 2 (1995): 77–93.

28. I. D. Steiner, *Group Processes and Productivity* (New York: Academic Press, 1972).

29. R. M. Stogdill, "Group Productivity, Drive, and Cohesiveness," *Organizational Behavior and Human Performance* 8 (1972): 26–43.

formance, but some have argued that these effects were often moderated by additional variables.[30] The current view is that cohesion does have some influence over team processes and how well a team performs, but that the effect often depends on the type of cohesion, the type of task, and the interaction with other team variables. In the case of group performance, a majority of authors cite task cohesion as the critical component in the cohesion-performance effect. For example, task cohesion has been related to better performance for teams making decisions under temporal stress.[31] Particularly for additive tasks, for which individual efforts are combined to complete an overall group task, S. J. Zaccaro and C. A. Lowe found that high task cohesion increased performance, but that interpersonal cohesion had no effect.[32]

Such is not the case with disjunctive tasks, for which group members must work together to produce a collective product. Zaccaro and M. C. McCoy had groups rank 15 items in order of importance to group survival in a simulated survival situation task. Results indicated that high task and high interpersonal cohesion groups outperformed groups either high on one type but low on another, or low on both types. For disjunctive tasks, Zaccaro and McCoy noted, "High task-based cohesion increases the likelihood that high ability members will contribute to the group problem-solving, whereas high interpersonal cohesion facilitates the procurement, recognition, and acceptance of high quality contributions."[33] In addition, better-performing teams competing in a complex business simulation game, a disjunctive-type task, were more cohesive, as represented by higher scores on measures of interpersonal and task cohesion. The authors maintained that cohesive teams "are better performers because they are able to satisfy the social needs of the team members while simultaneously demonstrating a shared commitment to the team task."[34]

---

30. S. M. Gully, D. J. Devine, and D. J. Whitney, "A Meta-analysis of Cohesion and Performance: Effects of Levels of Analysis and Task Interdependence," *Small Group Research* 26, no. 4 (1995): 497–520.

31. Zaccaro, Gualtieri, and Minionis, "Task Cohesion": 77–93.

32. S. J. Zaccaro and C. A. Lowe, "Cohesiveness and Performance on an Additive Task: Evidence for Multidimensionality," *Journal of Social Psychology* 128, no. 4 (1988): 547–558.

33. S. J. Zaccaro and M. C. McCoy, "The Effects of Task and Interpersonal Cohesiveness on Performance of a Disjunctive Group Task," *Journal of Applied Social Psychology* 18, no. 10 (1988): 837–851.

34. P. Miesing and J. Preble, "Group Processes and Performance in a Complex Business Simulation," *Small Group Behavior* 16 (1985): 325–338.

Despite the apparent support that task cohesion and, to a lesser extent, interpersonal cohesion positively influence team and group performance, part of the difficulty in defining the cohesion-performance effect is attributed to the influence of moderator variables. C. R. Evans and K. Dion, in their meta-analyses of over 372 groups, showed that group cohesion led to increased performance; however, the effect was relatively small and appeared to depend on other factors.[35] One example is A. Tziner and Y. Vardi's finding that for three-person tank crews, performance, effectiveness, and cohesiveness were correlated only when studied in combination with the command style of tank commanders. Highly cohesive teams exhibited better performance only if the command style emphasized an orientation toward the task and the team members. For command styles only emphasizing team member orientation, low cohesiveness was related to better performance.[36] Another perspective is that cohesion is associated with performance, but that high levels of team cohesion may negatively affect a team, as in the case of groupthink or in teams whose norms do not support productivity.[37] If the predominate group norm is a slow work pace, cohesiveness might actually reduce performance.

Similar to the effect of moderator variables, another complicating factor in studying the cohesion-performance effect is determining which comes first. There remains significant controversy over the causal nature of the relationship. In the sports domain, Daniel Landers, Michael Wilkinson, Brad Hatfield, and Heather Barber commented, "Even when the same measuring instruments are employed for interacting team sports, some studies demonstrate a reciprocal causality between the two variables (i.e., cohesion affects performance outcome and vice versa), whereas other studies find that performance outcome affects cohesion, but cohesion does not influence performance."[38]

---

35. C. R. Evans and K. Dion, "Group Cohesion and Performance: A Meta-analysis," *Small Group Research* 22, no. 2 (1991):175–186.

36. A. Tziner and Y. Vardi, "Ability as a Moderator Between Cohesiveness and Tank Crew's Performance," *Journal of Occupational Behavior* 4 (1983): 137–143.

37. T. W. Porter and B. S. Lilly, "The Effects of Conflict, Trust, and Task Commitment on Project Team Performance," *International Journal of Conflict Management* 7, no. 4 (1996): 361–376.

38. Daniel M. Landers, Michael O. Wilkinson, Brad D. Hatfield, and Heather Barber, "Causality and the Cohesion-Performance Relationship," *Journal of Sport Psychology* 4, no. 2 (1982): 170–183.

Taken together, the above studies suggest that there is a relationship between the cohesion of a team or crew and its performance, although the specifics remain unclear. In the context of LDSF, it is even more difficult to research the construct given that crew cohesion is not stable over the course of the mission. Dion, for example, cites several studies showing that cohesion declines in the middle and later stages of a mission as reflected by increases in crew tension and conflict.[39] Likewise, Nick Kanas found that cohesion levels were significantly higher during the first few weeks of missions than in later stages.[40]

In summary, despite methodological differences between cohesion-performance studies, the influence of moderator variables, and disagreement over the direction of the relationship, several conclusions are possible. First, both task and interpersonal cohesion may improve performance and group processes, but task cohesion more consistently predicts performance. Second, on additive-type tasks, high interpersonal cohesion can have a negative effect on performance due to more non-task-relevant conversations between team members, but high levels of both task and interpersonal cohesion benefit performance on disjunctive tasks. What these studies do not demonstrate, however, is how the composition of the team with regard to gender affects the development and maintenance of cohesion.

GENDER COMPOSITION AND CREW COHESION

With regard to gender specifically, the team literature suggests that men and women do work in slightly different ways, such as in their leadership styles and reactions to stress, that can influence cohesion. In addition, even though the unique contributions from each gender often improve team performance in some extreme environments analogous to spaceflight, there are concerns over how gender stereotypes and disagreements negatively affect crew interactions. If we assume that teams possessing higher levels of cohesion, in general, perform more effectively than teams with low cohesion levels, then a primary consideration when selecting

---

39. Dion, "Interpersonal and Group Processes in Long-Term Spaceflight Crews": C39.
40. N. Kanas, "Group Interactions in Space," *Aviation, Space, and Environmental Medicine* 75, no. 7, sect. II (2004): C4.

crews for LDSF is defining the optimal combination of individuals to maximize cohesion. In addition to size, experience, and culture, an important question is what mixture of men and women is ideal for a long-duration space mission? Preferably, we would rely on results from empirical studies on gender and cohesion; however, few researchers have addressed this relationship specifically, fewer still in the context of LDSF. Nevertheless, if we focus first on team performance in general, findings from the business and military domains offer some insight into the effects of gender. For example, manufacturing teams with a larger number of women taking on informal leadership roles within the team received higher supervisor ratings than teams with fewer female leaders.[41] Likewise, as women were added to technical teams at a Fortune 500 aerospace company in a stepwise fashion, the addition of one or two women did not adversely affect team performance in comparison to that of all-male teams, and adding three or four women produced a slightly positive effect on team performance.[42] In contrast, some studies suggest that gender heterogeneity is problematic for teams, at least in terms of ratings of team effectiveness, particularly when there is diversity in composition variables beyond gender, like ethnic and cultural background. Gayle Baugh and George Graen found that members of project teams in a state regulatory agency rated their teams as less effective when the members varied in terms of gender and race than when the team members were all male or all white. However, ratings of the teams by outside evaluators showed no differences with regard to gender and race.[43]

As already described, there are few studies focused exclusively on gender and cohesion; however, available research does suggest that the inclusion of women on a team, at the very least, does not negatively affect cohesion and in some cases actually improves cohesion. In a military context, arguably a better analog to spaceflight than business, a Women in Combat Task Force Study Group concluded that

---

41. Mitchell J. Neubert, "Too Much of a Good Thing or the More the Merrier? Exploring the Dispersion and Gender Composition of Informal Leadership in Manufacturing Teams," *Small Group Research* 30, no. 5 (1995): 635–646.

42. Janet W. Burris, "The Impact of Gender Diversity on Technical Team Effectiveness," *Dissertation Abstracts International: Section B: The Sciences and Engineering* 62, no. 10-B (May 2002): 4715.

43. Gayle S. Baugh and George B. Graen, "Effects of Team Gender and Racial Composition on Perceptions of Team Performance in Cross-functional Teams," *Group and Organization Management* 22, no. 3 (1997): 366–383.

women had either a positive or a neutral effect on the type of cohesion present in military units. This model of cohesion takes into account team factors such as interdependence, unit identity, personnel stability, communications, and leadership.[44] Similarly, Robert Vecchio and Donna Brazil's survey of nearly 2,000 U.S. armed services cadets indicated that increases in the number of women in a squad were not associated with any decreases in squad-level measures of cohesion.[45] In the aviation domain, an even better analog to spaceflight, the relationship between gender and cohesion is less clear. For example, four-person teams of students from a Florida university, flying an F-22 flight simulation, exhibited higher levels of interpersonal cohesion when the teams were of the same gender. Furthermore, this increased interpersonal cohesion helped to enhance coordination between team members, leading to improved performance.[46]

Clearly, additional empirical investigations are warranted to more completely define how gender heterogeneity in teams affects cohesion. In lieu of controlled research, a tenable approach is to survey personnel from space and analog settings, although very few studies using this approach have specifically focused on cohesion. For example, according to Rosnet and colleagues, including women in wintering groups at polar stations "seems to have positive effects on the general climate of the group by partly limiting men's rude behavior, but it also seems to be an important stressor for both men and women when the females' average age is close to the males'."[47] Similarly, a former commander of the U.S. Naval Support Force Antarctica, who had experienced both all-male winter-over missions and gender-integrated stays, reported that women had a stabilizing effect on personnel and believed these heterogeneous groups were more productive than all-male groups.[48] In space, an American woman who lived aboard the ISS reported that the crew

---

44. V. J. Saimons, "Women in Combat: Are the Risks to Combat Effectiveness Too Great?" Monograph Report No. AD A258 247 (Fort Leavenworth, KS: U.S. Army Command and General Staff College, School of Advanced Military Studies, 1992).

45. Robert P. Vecchio and Donna M. Brazil, "Leadership and Sex-Similarity: A Comparison in a Military Setting," *Personnel Psychology* 60, no. 2 (2007): 303–335.

46. Frederick-Jorge Panzer, "The Influence of Gender and Ethnic Diversity on Team Effectiveness," *Dissertation Abstracts International: Section B: The Sciences and Engineering* 64, no. 3-B (2003): 1534.

47. Rosnet et al., "Mixed-gender Groups": C12.

48. Stuster, *Bold Endeavors*, p. 178.

interacted well and achieved all mission goals.[49] On the other hand, reports from space analogs such as offshore oil rigs, naval vessels, and Antarctic bases indicate that interpersonal problems related to mixed-gender crews had a negative effect on crew performance.[50] Stuster notes that "on closer inspection, however, the problems appear to have been not directly attributable to mixed crews, but rather to the behavioral consequences of immaturity, faulty personnel selection, and inadequate pre-mission training for both male and female members of the crews."[51]

Another approach is to consider how gender heterogeneity affects factors closely related to the development of cohesion. Recall that Bishop's conclusion after reviewing literature from extreme environments was that the presence of women in mixed-gender crews appeared to promote crew behaviors that improved its functioning.[52] Likewise, Endler's finding that women employ a more interpersonal and caring approach when dealing with stress may significantly improve the interpersonal atmosphere within a crew, improving interpersonal cohesion by bringing members closer together.[53]

CONCLUSION

The success of future human space missions rests squarely on the shoulders of the men and women who will venture into space for months, possibly years, at a time. In this chapter, we argued that despite inherent differences in the behaviors and abilities of men and women, mixed-gender crews have performed effectively, both in space and in similar settings like Antarctica. In most cases, teams composed of both men and women function as well as or better than all-male teams. We also supported the conclusion that crew interactions, specifically the level of cohesion within the crew, are extremely important to the crew's overall performance. Available evidence, albeit limited in scope and size, indicates that

---

49. Kanas and Manzey, *Space Psychology and Psychiatry*, p. 77.
50. Stuster, *Bold Endeavors*, p. 177.
51. Ibid.
52. Bishop, "Evaluating Teams in Extreme Environments": C17.
53. Endler, "The Joint Effects of Person and Situation Factors": C25.

gender-homogenous crews exhibit similar, and at times higher, levels of cohesion than gender-heterogeneous crews in settings similar to spaceflight.

In this chapter, we concentrated on just one piece of this puzzle related to the gender composition of the crew and how the mixture of men and women affects the crew's task and interpersonal cohesion. Our conclusion, based on available research from space, space analogs, and other team-related activities, is that a crew composed of both women and men is the right choice for extended missions to the International Space Station, rendezvous missions with asteroids, and, one day, the first human mission to Mars.

Chapter 7

Flying with Strangers: Postmission Reflections of Multinational Space Crews

Peter Suedfeld[1]
   Department of Psychology
   University of British Columbia

Kasia E. Wilk
   Youth Forensic Psychiatric Services Research and Evaluation Department
   Ministry of Children and Family Development

Lindi Cassel
   Department of Occupational Therapy
   Providence Health Care

ABSTRACT

After the Space Age began as part of the national rivalry between the USSR and the United States, space exploration gradually took on a multinational character as both countries included astronauts from their respective allies, and eventually from each other, in their missions. This trend became institutionalized in the Shuttle-*Mir* program and in the construction of the International Space Station (ISS). The latter is the first truly international, as opposed to multinational, space capsule, in that it does not belong to and was not built by one country. In previous cases, one national space agency was always the host and crewmembers from other nations were perceived and treated as guests. This "guest" status, which usually

---

1. This research was made possible by Contract No. 9F007-033006 with the Canadian Space Agency and is part of the project Long-term Effects After Prolonged Spaceflight (LEAPS). A briefer version of the chapter was presented at the meeting of the American Psychological Association in San Francisco, CA, in August 2007. Correspondence should be addressed to Peter Suedfeld, Department of Psychology, University of British Columbia, Vancouver, BC V6T 1Z4, Canada, or *psuedfeld@psych.ubc.ca*.

went with being a minority among a majority from the "host" nation, led to considerable dissatisfaction and frustration.

This chapter examines the archived reminiscences of both majority and minority astronauts and cosmonauts, relying primarily on the method of Thematic Content Analysis (TCA). TCA is a set of techniques whereby trained scorers identify and quantify specific variables in narratives. In this study, TCA procedures were used to analyze how majority-minority status and other variables (e.g., gender, mission duration, and Space Age era) affected satisfaction, feelings about crewmates and home agencies, personal values, ways of coping with problems, and other psychosocial reactions of the mission participants. The study drew upon astronauts' and cosmonauts' memoirs, autobiographies, media interviews, and oral history interviews as the databases on which TCA scoring was performed.

NATIONALISTIC EMBODIMENTS OF A UNIVERSAL HUMAN DRIVE

The exploration of space may be attributed to two driving forces. One is an innate drive shared by many species but perhaps best exemplified by humanity: the urge to seek novelty, to enlarge the sphere of the known as we advance into the hitherto unknown, and to expand the habitat of humankind.[2] Long before technology made real space voyages possible, fictional explorations can be traced to the myth of Daedalus and Icarus and its counterparts in other traditions, to the writings of Cyrano de Bergerac, and eventually to the imaginations of Jules Verne and the multitude of early-20th-century science fiction writers.

The second motivator, which determined just when in our species' history space travel would move from fiction to reality, was international rivalry. Primitive military rocketry began centuries ago, accelerated and took the first large steps toward space during World War II, and was increasingly well supported and brought to eventual fruition as the "space race" component of the Cold War.

---

2. M. Holquist, "The Philosophical Bases of Soviet Space Exploration," *The Key Reporter* 51, no. 2 (winter 1985–86): 2–4.

Flying with Strangers: Postmission Reflections of Multinational Space Crews

The first decades of human spaceflight were a series of competitions between the Soviet Union and the United States: who would be the first to launch an orbiting spacecraft, a piloted spacecraft, a space crew, a Moon rocket, a space station . . . . Flights were scheduled to preempt media publicity from the competition. Temporary victory veered from one bloc to the other, with each claiming—or at least implying—that being momentarily ahead in the race was proof of the superiority of its political and economic system, just as Olympic gold medals were (and are) risibly interpreted as markers of national quality.

In such a setting, it followed logically that cooperation between the two leading space nations would be unlikely. The original space travelers were exemplars of the virtues each country extolled: they were military pilots, the cream of that already hand-picked crop, who were used to flying experimental and operational aircraft at the very edge of new technology, individuals of demonstrated courage, coolness, and ability. The world was shown that they were all physically fit, psychologically stable, good husbands and fathers, modest, humorous, and loyal. They were patriotic citizens and, depending on which program they were in, strong supporters of either communism or democratic capitalism. Although these portraits omitted a number of what would have been more realistic, if less rosy, individual differences among these pioneers, both space agencies continued to paint such idealized pictures, and the spacemen did their best not to smear the paint (although later in the Space Age, revisionists have tried to rub off some of its luster by emphasizing the internal politics of the agencies, alleging arbitrary and biased decisions being made concerning the assignment of astronauts, and so on).[3] More recently, selection procedures have changed to reflect the expanded sources and duties of astronauts, to include civilians, nonpilots, women, and a variety of (mostly, but not entirely, technical and scientific) professionals; but there is a perception that some kinds of bias still exist—e.g., in favor of astronauts from the military.[4]

It is worth remembering that the combination of the universal urge to explore and the particularistic urge to use exploration to exalt one's nation is neither new nor unique to space explorers. For centuries, it has been a prominent reason why

---

3. B. Burrough, *Dragonfly: NASA and the Crisis Aboard Mir* (New York: HarperCollins, 1998).
4. M. Mullane, *Riding Rockets: The Outrageous Tales of a Space Shuttle Astronaut* (New York: Scribner, 2006).

terrestrial expeditions were funded and also a strong component of many expeditioners' motivation.[5]

GUEST ROOMS IN SPACE

In 1975, the two rivals cooperated to design a docking module that allowed spacecraft from each (Apollo and Soyuz) to join in space. Later, both superpowers began to offer room and board in their space capsules to citizens of their respective international blocs. The Soviet *Interkosmos* program made room for cosmonauts from various Eastern Bloc countries, as well as from France, Syria, and India; American crews have shared their spacecraft with colleagues from Canada, Western Europe, Australia, Japan, India, Israel, Brazil, and Saudi Arabia.

This trend was reinforced by the establishment of space agencies in countries that could select and train astronauts but had no independent crewed space vehicles. The most active among these are Canada, Germany, France (and eventually the European Union [EU]), and Japan. The People's Republic of China has since gone beyond such strategies to develop its own launch vehicles and begin an independent program of piloted spaceflight. Eventually, multinationality became routine, as did the inclusion of women and the broadening of selection to allow for the participation of people who were not military, not test pilots, and often not even pilots. The new participants were from a range of disciplines: engineers, scientists, physicians, politicians, and, most recently, private individuals who bought a brief stay on the ISS.

This major increase in the diversity of space voyagers sharpens a distinction that began when the USSR and the United States first added foreign crewmembers. Differences, sometimes invidious, were not only between nationalities per se, but also between the "host" crew of Americans or Soviets/Russians and the "visitors." At first, the inclusion of international crewmembers was primarily a propaganda move. It had relatively little beneficial effect on the missions themselves and angered the established astronaut and cosmonaut corps by reducing the flight

---

5. J. R. L. Anderson, *The Ulysses Factor: The Exploring Instinct in Man* (New York: Harcourt Brace Jovanovich, 1970).

Flying with Strangers: Postmission Reflections of Multinational Space Crews

opportunities of their members.[6] Until the construction of the ISS, every capsule that carried human beings into space was either American or Soviet/Russian. Was it possible that mixed-nationality crews aboard felt equally at home and comfortable, or was a host-guest distinction unavoidable? Would the latter be strengthened by the fact that some of the "home" team inhabited the vehicle for a much longer period than did the foreign visitors? Could the distinction be eliminated, or at least minimized, by appropriate training and crew composition? And what did the answers to these questions imply for truly international efforts, such as building and working on the ISS, and perhaps the eventual exploration of space beyond low-Earth orbit and the Moon? This chapter presents data that address some, though not all, of these questions.

There has been considerable evidence that psychosocial stressors are among the most important impediments to optimal crew morale and performance.[7] Positive reactions during and after spaceflight were relatively ignored as psychologists focused on problems that needed to be avoided or solved. After a somewhat slow start toward balancing the situation, attention to positive aspects has expanded in the past few years to look at eustress (positive stress), personal growth, excitement, enjoyment, feelings of satisfaction, camaraderie, and changes in values.[8]

---

6. T. Furniss and D. J. Shayler, with M. D. Shayler, *Praxis Manned Spaceflight Log, 1961–2006* (Chichester, U.K.: Springer Praxis, 2007); Mullane, *Riding Rockets*.

7. N. Kanas and D. Manzey, *Space Psychology and Psychiatry* (Dordrecht, Netherlands: Kluwer, 2003); Space Studies Board, *A Strategy for Research in Space Biology and Medicine in the New Century* (Washington, DC: National Research Council, 1998); J. Stuster, *Bold Endeavors: Lessons from Space and Polar Exploration* (Annapolis, MD: Naval Institute Press, 1996); P. Suedfeld, "Applying Positive Psychology in the Study of Extreme Environments," *Journal of Human Performance in Extreme Environments* 6 (2001): 21–25; P. Suedfeld, "Space Memoirs: Value Hierarchies Before and After Missions—A Pilot Study," *Acta Astronautica* 58 (2006): 583–586.

8. Suedfeld, "Applying Positive Psychology": 21–25; E. C. Ihle, J. B. Ritsher, and N. Kanas, "Positive Psychological Outcomes of Spaceflight: An Empirical Study," *Aviation, Space, and Environmental Medicine* 77 (2006): 93–102; A. D. Kelly and N. Kanas, "Communication Between Space Crews and Ground Personnel: A Survey of Astronauts and Cosmonauts," *Aviation, Space, and Environmental Medicine* 9 (1993): 795–800; P. Suedfeld, "Invulnerability, Coping, Salutogenesis, Integration: Four Phases of Space Psychology," *Aviation, Space, and Environmental Medicine* 76 (2005): B61–B66; Suedfeld, "Space Memoirs": 583–586; P. Suedfeld and G. D. Steel, "The Environmental Psychology of Capsule Habitats," *Annual Review of Psychology* 51 (2000): 227–253.

It has been pointed out that "mixed" crews are mixed in many different ways. Intercultural issues can arise, and have arisen, not only between space voyagers of different nationalities, but also between those of different space agencies, sexes, and educational and professional backgrounds. Crewmembers who came to space with a military test pilot background and those with an academic science background may have problems understanding each other's jargon and worldview (to say nothing of those of teachers and politicians). The same, to an even greater extent, is likely to be true in international crews that are not perfectly bilingual.[9] However, the current chapter focuses on only one kind of diversity, that based on nationality.

Whether the possible benefits of increased diversity in crew composition (such as reducing boredom, celebrating unaccustomed holidays, and becoming acquainted with new and useful approaches to interpersonal and operational problems) will outweigh the additional stresses that it generates, or vice versa, needs to be assessed through empirical data. To date, there have been three sources of relevant information. One advantage that they all share, which sets them off from simulation and analog studies, is their high external validity: the information is produced by real participants in real space operations. This is the only kind of information that will be considered here.

The most colorful and memorable, but least generalizable and scientifically rigorous, source is the collection of anecdotes that has been generated by the space voyagers and others involved in the programs. Self-report studies using surveys and interviews have provided both qualitative and some quantitative information, usually from a relatively small number of crewmembers during a mission and occasionally from larger samples of ground staff personnel. Thematic content analyses applied to interviews, memoirs, and similar archival materials provide another form of quantitative analysis applied to qualitative materials. This is the method used in the current chapter.

---

9. Kelly and Kanas, "Communication Between Space Crews and Ground Personnel": 795–800; P. Kumar, "Intercultural Interactions Among Long-Duration Spaceflight Crew (LDSF)" (paper presented at the International Astronautical Congress, Hyderabad, India, September 2007).

Flying with Strangers: Postmission Reflections of Multinational Space Crews

"MY HOUSE" OR JOINT TENANCY?
Anecdotal Evidence

As in much of space psychology, and more generally in the psychology of all extreme and unusual environments, the first bits of knowledge came from the anecdotes told and written down by participants. These stories have tended to emphasize the dramatic, and therefore mostly unpleasant, interactions between crewmembers of different demographic (including national/cultural) categories. It should be noted that most of them are "common currency" in the space community; the references given are only examples of several sources in which these stories have appeared.

The kind of diversity with which this paper is concerned, that is, differences in national origin, has been the topic of many anecdotal reports. Some of the best known involve visitors to Soviet capsules. When the first *Interkosmos* cosmonaut, Vladimir Remek of Czechoslovakia, returned from space (Soyuz 28, 1978), the joke went around that he was suffering from "red hand syndrome": every time he reached for a switch or other control, a Russian crewmate would slap his hand and tell him not to touch it.[10]

Four years later, Jean-Loup Chrétien, a French air force officer and the first of a series of French cosmonauts, was likewise forbidden to touch anything during his crew training with two Russians; he not-so-subtly communicated his frustration (and annoyed his crew commander) by bringing a pillow and going to sleep during one training session. After the inhospitable commander was replaced and Chrétien reached the *Salyut* space station for a one-week visit, his expertise, good nature, and sophisticated equipment impressed the Russians—but one of them later expressed his relief at going back to black bread and borscht after a menu of canned French delicacies, including compote of pigeon with dates and dried raisins, duck with artichokes, boeuf bourguignon, and more.[11] Chrétien, in turn, criticized the excessive workload imposed on the crew.[12]

---

10. V. Lebedev, 1990, cited in R. Zimmerman, *Leaving Earth: Space Stations, Rival Superpowers, and the Quest for Interplanetary Travel* (Washington, DC: Joseph Henry Press, 2003), p. 134.

11. V. Lebedev, *Diary of a Cosmonaut: 211 Days in Space* (New York: Bantam Books, 1990; original publication, 1983); "Surprise! Astronauts Eat in Orbit," *Space Today Online*, http://www.spacetoday.org/SpcShtls/AstronautsEat.html (accessed 14 March 2008).

12. R. D. Hall, D. J. Shayler, and B. Vis, *Russia's Cosmonauts: Inside the Yuri Gagarin Training Center* (Chichester, U.K.: Springer Praxis, 2005), pp. 235–236.

The long-duration deployments to the *Salyut* and *Mir* space stations included the presence of mixed crews, and the Shuttle-*Mir* mission series was in fact designed for such crews. Each of the latter missions was constructed around an American-Russian team flying to *Mir* aboard a Space Shuttle orbiter and remaining on the station (with occasional crew changes and short-term visitors) for between four and seven months.

The reluctance of Russian hosts to admit their guests to full coworker status persisted during this collaborative program. In 1995, Norman Thagard was the first American to be a long-term crewmember on the *Mir* space station. Despite his status as a full resident, rather than a short-term visitor like Remek and Chrétien, Thagard, like them, felt that he was left out of important and interesting activities on the aging and deteriorating spacecraft. He wound up doing crossword puzzles while his crewmates did the work. Shannon Lucid, who spent six seemingly happy months on *Mir* in 1996, was left "in command" of the station while her two Russian colleagues performed EVAs; however, the control switches were taped down, and she was told not to touch anything.[13] In an oral history interview, one NASA psychologist said, "We were never able, I don't think, to have the American be on par with the Russian crew members . . . ."[14] The problem may not be restricted to the astronauts. Thagard and other Shuttle-*Mir* astronauts indicated that more vigorous support from NASA ground personnel in Mission Control in Russia might have ameliorated these problems—but those personnel in turn felt themselves to be tense, unhappy, underutilized, and somewhat ignored by their own Russian counterparts.[15]

---

13. S. Lucid, "Six Months on *Mir*," *Scientific American* (May 1988): 46–55; Zimmerman, *Leaving Earth*.

14. Al Holland, interview by Rebecca Wright, Frank Tarazona, and Summer Bergen, 13 August 1998, published through "Shuttle-Mir Oral History Project," Johnson Space Center History Portal, available at *http://www.jsc.nasa.gov/history/oral_histories/participants.htm* (accessed 7 June 2010).

15. J. M. Linenger, *Off the Planet: Surviving Five Perilous Months Aboard the Space Station* Mir (New York: McGraw-Hill, 2000); Norman E. Thagard, interview by Rebecca Wright, Paul Rollins, and Carol Butler, 16 September 1998, published through "Shuttle-Mir Oral History Project," Johnson Space Center History Portal, available at *http://www.jsc.nasa.gov/history/oral_histories/participants.htm* (accessed 5 May 2007); Zimmerman, *Leaving Earth*; N. Kanas, V. Salnitskiy, E. M. Grund, et al., "Interpersonal and Cultural Issues Involving Crews and Ground Personnel During Shuttle/Mir Space Missions," *Aviation, Space, and Environmental Medicine* 71, no. 9 (2000): A11–A16.

Even those astronauts who were given work to do could wind up with menial or routine jobs.[16] David Wolf, a Shuttle-*Mir* resident astronaut, volunteered to clean "gooey, slimy, ice-cold fluid" from the station's walls, a job that then devolved on him for 4 to 8 hours per day, almost every day, while his Russian colleagues performed sophisticated technical work. Wolf accepted this with equanimity: "that was the best thing I could come up with to free up their time for what they're better at and be part of the team."[17]

The critical attitude toward people perceived to be not-quite-colleagues was not restricted to the Russian space program. Mike Mullane, referring to "part-time astronauts"—one-flight foreign visitors, payload specialists, politicians, and the like—asserts that their training had been cursory and superficial, that some of them exhibited psychological problems, and that "Mission commanders provided their own additional training in the form of the admonishment 'Don't touch any shuttle switches!'"[18] Obviously, "part-time astronauts" were seen as less expert and therefore undependable. J. M. Linenger, too, comments negatively on his and colleagues' attitude toward *American* "part-time astronauts."[19]

Of course, this should not have applied to people such as Remek, Thagard, and Lucid. They and many others who flew as national minorities were in fact professional astronauts. They were just as well trained as the national majority with whom they flew, and in many cases, they trained together with the majority for a year or more. The comments of majority crewmembers are typically quite positive about their foreign colleagues' personality and ability to get along with the rest of the crew, but the distrust in their competence within the "home team's" spacecraft (and/or with the home team's language) persisted nonetheless.[20]

---

16. N. Kanas, V. P. Salnitskiy, E. M. Grund, V. I. Gushin, D. S. Weiss, O. Kozerenko, A. Sled, and C. R. Marmar, "Social and Cultural Issues During Space Missions," *Acta Astronautica* 47 (2000): 647–655.

17. David Wolf, interview by Rebecca Wright, Paul Rollins, and Mark Davison, 23 June 1998, "Shuttle-Mir Oral History Project," Johnson Space Center History Portal, available at *http://www.jsc.nasa.gov/history/oral_histories/participants.htm* (accessed 7 June 2010).

18. Mullane, *Riding Rockets*.

19. Linenger, *Off the Planet*.

20. N. Thagard, interview with the Panel on Human Behavior, Space Studies Board, National Research Council, Washington, DC, 2 May 1997.

These and similar stories may not be representative of the general experiences of national minorities in a space crew. Many of these individuals' recollections were primarily positive. Nevertheless, the negatively toned anecdotes point out, even if they may overemphasize, problems of which planners should be aware.

To some extent, friction between majority and minority crewmembers may arise from differences in the home cultures of the two most populous groups, Russians (including citizens of the former USSR) and Americans. J. B. Ritsher, in an excellent summary of the relevant cultural differences between these two nations, cites research not only from space but also from aviation and from psychological, sociological, and anthropological studies more generally.[21] According to these studies, Russian culture values collectivism, hierarchical power, distance, and paternalism more than American culture and values individualism, egalitarianism, mastery, autonomy, and uncertainty less. In addition, the USSR was oriented more toward survival and less toward well-being.

On a number of dimensions, including the ones listed above, Russia is discrepant from all of the other nations involved in the International Space Station project. Russian and American approaches to spaceflight differ in significant ways, some of these reflecting the more general cultural differences discussed by Ritsher. Supporting the view that Russian culture is more hierarchical than American culture were perceptions that Russian mission commanders were more authoritarian, Russian communications were more structured and formal (at least in English translation), and Russians were more circumspect in criticizing mission control or the systems on the spacecraft.[22]

Perhaps because of differences in the national economies and the funding of the space agencies, cosmonauts were more likely to feel that they had to try to repair malfunctioning systems, whereas astronauts tended to discard and replace them. Russians consequently were more satisfied with systems that worked adequately rather than demanding that they work perfectly. On a more personal level, cosmonauts (unlike astronauts) are paid a large spaceflight bonus, with deductions based on how many of the preset goals (experiments, repairs, extravehicular activities, etc.) they fail to complete successfully during the mission. As a result, their

---

21. J. B. Ritsher, "Cultural Factors and the International Space Station," *Aviation, Space, and Environmental Medicine* 76 (2005): 135–144.

22. Kumar, "Intercultural Interactions."

foreign colleagues sometimes considered them to be reluctant to do work that was not within the pay-for-action agreement and to do anything that might foil the completion of an agreed-upon paid task.[23]

Although it is certainly likely that cultural "root causes" (especially those stemming from differences between Russian and Western cultures) may underlie some of the frictions between majority and minority crewmembers, the host-guest dichotomy may have caused more problems than cultural or national diversity per se. If so, a completely different picture may emerge within a truly international facility such as the ISS when it becomes fully operational.

There is another possible explanation. Valentin Lebedev, a long-duration *Mir* cosmonaut, recognized a difference between his reactions to foreign and to compatriot visitors. Concerning one of the former, he wrote, "It's nice to have guests, but they make you tired," even though most of his comments about his French colleague were positive; commenting on an upcoming visit by fellow Russian cosmonauts, he wrote, "I think it will be easier with this visiting crew; they won't disturb us as much . . . ."[24] It may be that it is not nationality but familiarity that makes a visitor more welcome, so that more extensive pre-mission training and joint activities might erase or at least diminish the invidious difference.

Self-Report Studies

Kanas and Manzey summarized the few studies using self-report measures by space voyagers who had flown in foreign company.[25] Although there have been several simulation and analog studies (respectively, group isolation experiments in specially designed settings and field studies in isolated areas such as the polar regions and undersea habitats), data from actual spaceflight are scarce. Participants have reported miscommunications due to both spoken and nonverbal interaction styles, abrasive differences in leadership decision-making, differences in work patterns,

---

23. Ibid.
24. Lebedev, *Diary of a Cosmonaut*, pp. 101 and 189, respectively.
25. Kanas and Manzey, *Space Psychology and Psychiatry*.

different standards of hygiene and food preparation, and personality clashes that may be related to cultural factors.

Kanas and his colleagues have conducted major studies of space crews in flight by administering standard questionnaires that crewmembers can complete on a computer while the mission is going on. One such study, of crewmembers on Mir (five Americans and eight Russians) and the ISS (eight and nine, respectively), found that cosmonauts on Mir experienced more direction, support from the leader, and self-discovery than astronauts; lower vigor and more tension and anxiety on the ISS; and less job pressure but higher task orientation and managerial control in both places.[26] Americans on the Russian station felt less comfortable and less well supported from the ground than did the "home team." In contrast, ISS procedures are more U.S.-influenced, which may have made the Russians feel that they were on unfamiliar territory. Another report on the same Shuttle-Mir crews found that during the second half of the mission, Russian crewmembers reported decreasing cohesion and work pressure compared to Americans.[27]

A more recent study reported that miscommunications abounded when members of international crews engaged in extravehicular activities, but not when all crewmembers were from the same country, and that besides the obvious language barrier, space fliers generally felt that coming from the same cultural background would also reduce interpersonal friction.[28] Most of those interviewed agreed that on long-duration missions, they would prefer to go with a homogeneous crew from their own culture. In fact, according to Linenger, many U.S. astronauts declined the opportunity to participate in the Shuttle-Mir program for reasons that included distrust of Russian technology and post–Cold War hostility toward Russians themselves. Some of those who did agree to join the program were so dissatisfied with the training they got in Russia that they threatened to quit—"a near mutiny."[29]

---

26. N. Kanas, V. P. Salnitskiy, J. B. Ritsher, V. I. Gushin, D. S. Weiss, S. A. Saylor, O. P. Kozerenko, and C. R. Marmar, "Human Interactions in Space: ISS vs. Shuttle/Mir" (paper IAC-05-A1.5.02, presented at the International Astronautical Congress, Fukuoka, Japan, October 2005).

27. N. Kanas, V. P. Salnitskiy, D. S. Weiss, E. M. Grund, V. I. Gushin, O. Kozerenko, A. Sled, A. Bosrom, and C. R. Marmar, "Crewmember and Ground Personnel Interactions over Time During Shuttle/Mir Space Missions," *Aviation, Space, and Environmental Medicine* 72 (2001): 453–461.

28. Kumar, "Intercultural Interactions."

29. Linenger, *Off the Planet*, p. 45.

Flying with Strangers: Postmission Reflections of Multinational Space Crews

THE CURRENT STUDY: THEMATIC CONTENT ANALYSIS

Content analysis is a research method used in many disciplines to study narratives of interest. For example, anthropologists may content-analyze myths or folktales to identify important issues or beliefs of a culture; literary scholars may find in novels or plays the dominant patterns of social relations in a particular time or place, or hints as to the childhood experiences and personality of an author. Such qualitative or impressionistic methods are frequently used to explore hypotheses derived from a particular theory such as psychoanalysis, Marxism, or postmodernism. The scholar finds examples in the material that are relevant to the theory and uses those examples as evidence, as in Freud's inferences about Leonardo da Vinci's family background, childhood, and personality, based primarily on the analysis of a dream that Leonardo recorded in his diary.[30]

A purely quantitative counterpart is computerized content analysis, in which the occurrence of certain kinds of words or phrases is counted and used to infer either historical or personal characteristics. For example, a frequent appearance of the word "I" may indicate a degree of self-confidence, independence, or narcissism; hostile terminology ("enemy," "threatening," "evil") reveals a bellicose emotional or cultural state: an increase in such words when referring to another person or country may be one indicator of a forthcoming confrontation.[31]

The method used in our study, Thematic Content Analysis (TCA), is different from both the qualitative or impressionistic approaches and purely quantitative computerized ones. In TCA, either all available material or a randomly selected subset is used so that the researcher's theoretical preconceptions cannot bias the selection of material to be analyzed; identifying information is removed as far as possible, as a safeguard against bias in the actual scoring; and the material is scored blindly by a qualified scorer using a detailed scoring manual to further reduce the chances of bias and of idiosyncratic scoring criteria. Generally, at least one other

---

30. S. Freud, *Leonardo da Vinci and a Memory of his Childhood* (New York: Norton, 1961; original publication, 1910).

31. R. C. North, O. R. Holsti, M. G. Zaninovich, and D. A. Zinnes, *Content Analysis: A Handbook with Applications for the Study of International Crisis* (Evanston, IL: Northwestern University Press, 1963).

scorer analyzes a percentage of the same passages to ensure interscorer reliability. Thus, from a qualitative (e.g., narrative) database, quantitative data are generated in a scientifically rigorous way and statistical analyses are made possible.[32]

The issue of accuracy always arises in retrospective materials, usually from two perspectives. One is the exactness of memory; the other is the possibility of impression management. In the current study, precision probably varied as a function of time since the experience (among other variables), which itself varied from very little, as in the *Life* magazine interviews of the first Mercury astronauts, to years in the case of book-length memoirs. In any case, the question of how precisely the narrators remembered events is not of critical importance to this study: we were not interested in compiling a history of their experiences, but rather in the emotions and motives that were associated with the events and that emerged during recall. As for impression management, although this is a likely mediating variable in any self-descriptive human narrative, the TCA scoring criteria are not very transparent, and the material includes a number of cross-checks (e.g., prepared versus spontaneous remarks). Many of the narratives included negative reflections on both other people and the narrator himself (or herself), and stories by several participants in the same event showed substantial differences, so at least the attempt to make oneself (or one's colleagues or one's agency) look good did not swamp all other considerations, and there was no evidence of externally imposed uniformity in the accounts.

Method

The current study applied TCA scoring methods to a collection of memoirs, interviews, and oral histories originated by 63 astronauts and cosmonauts. The overwhelming majority of U.S. and Soviet/Russian participants were in the categories that NASA considers professional astronauts: pilots and mission specialists. The few exceptions were "spaceflight participants": individuals flown for some goal such as public relations. No "space tourists" (i.e., individuals who were allowed

---

32. C. P. Smith, J. W. Atkinson, and D. C. McClelland, eds. *Motivation and Personality: Handbook of Thematic Content Analysis* (Cambridge: Cambridge University Press, 1992).

to fly for a brief visit to the ISS upon payment of a multi-million-dollar fee) were included; neither were payload specialists, who fly as members of nongovernmental institutions such as corporations or universities to carry out a specific task or experiment. Participants from countries other than the United States and the USSR/Russia were a more mixed group, which included both professional astronauts and others (many of them professional air force officers) who, after their one spaceflight, would have no long-term connection with a space program.

The collections covered the era of human spaceflight from the very first period through the construction of the ISS, but we omitted reports related to missions in which crews of only one nation were involved. Due to the extremely small sample available from ISS veterans, we also omitted those data from our analyses. With the increasing number of crewmembers who have served on the Station, this problem may be on the way to being solved.

Because the source materials of this study varied widely in length, all TCA results reported below are based on number of category mentions per page in the source. Not every subject had references to all of the dependent variables, so n's (sample sizes within each subcategory) varied from measure to measure.

Anecdotes and a numerical content analysis software program were used as secondary data.

Independent Variables

Table 1 shows the breakdown of the subjects by relevant demographic and spaceflight categories, which served as the independent variables. "National origin" refers to the country with which the source was identified in the space program. For example, some "U.S." astronauts were originally immigrants from elsewhere; however, they were selected, trained, and chosen to fly by NASA. "USSR/Russia" includes cosmonauts whose citizenship was Soviet during the existence of the USSR and those who were Russian afterward. The "Other" category includes astronauts who had been recruited and selected by the established space programs of other nations (e.g., Canada or Japan, as well as France, Germany, or other EU nations) and who flew with either the U.S. or the Soviet/Russian space program. *Interkosmos* crewmembers and their equivalents flying with NASA are also classified as "Other."

Table 1. Number of subjects by category.

National Origin	Sex		Flew as		Flew with	
	Male	Female	Minority	Majority	U.S.	USSR/Russia
U.S.	16	10	7	19	20	6
USSR/Russia	14	0	4	10	6	8
Other*	19	4	23	—	19	4
Total	49	14	34	29	45	18

* "Other" refers to crewmembers who are neither American nor Soviet/Russian. All of the subjects in this category flew as a minority with either American or Russian majorities.

In addition to these analyses, others were performed on disaggregations based on mission duration (two weeks or less versus either four or six months or more) and mission phase (portions of the narrative referring to pre-, in-, or postflight periods). In some instances, the n within one of these cells was too small for analysis, and those scores are omitted from this report.

Dependent Variables

The scoring categories applied to the materials were as follows:
1. **Value hierarchies:** S. H. Schwartz defined values as having five major aspects. According to him, values
    1. are concepts or beliefs,
    2. pertain to desirable end states of behaviors,
    3. transcend specific situations,
    4. guide selection or evaluation of behavior and events, and
    5. are ordered by relative importance.[33]

Eventually, Schwartz reported that 11 categories of values underlying stable, important life goals (see table 2) had been empirically shown to have cross-cultural generality and high reliability.

---

33. S. H. Schwartz, "Universals in the Content and Structure of Values: Theoretical Advances and Empirical Tests in 20 Countries," *Advances in Experimental Social Psychology* 25 (1992): 4.

Table 2. Value categories and definitions (alphabetical order).*

Value	Brief Definition
Achievement	Personal success through demonstrated competence according to social standards
Benevolence	Concern for close others in everyday interaction
Conformity	Inhibition of socially disruptive acts, impulses, or inclinations
Hedonism	Pleasure in satisfying organismic needs
Power	Social prestige/status, control over people and resources
Security	Safety, harmony, stability of society, relationships, and self
Self-Direction	Independent thought and action: choosing, creating, exploring
Spirituality	Meaning and harmony by transcending everyday reality
Stimulation	Excitement, novelty, challenge
Tradition	Respect for one's cultural/religious customs and ideas
Universalism	Understanding, appreciation, tolerance, and protection of the welfare of all people and of nature

* Adapted from Schwartz, "Universals in the Content and Structure of Values": 5–12.

One way to group the values is to distinguish between those that serve individual interests and those that serve collective interests: Achievement, Hedonism, and Self-Direction versus Conformity, Security, and Tradition. The emphasis on individual versus collective cultural values is generally thought to separate American and Soviet cultures; a comparison between the two groups of values across the fliers representing the two national space agencies can be an interesting way to check on this widespread view.

The value scores reported in the current paper reflect the number of times the source person mentioned experiencing, advancing, or identifying with values in that category per page or section of text in the source material.

2. **Social relations** (two subcategories):
    a. Affiliative Trust/Mistrust: positive, trusting relationships versus cynicism and negativity toward others.[34]
    b. Intimacy: Positive Intimacy is a measure of readiness or preference for warm, close, and communicative interaction with others.[35] In the current study, we supplemented this by scoring Negative Intimacy as well: negative affect in relationships, negative dialogue, rejection of commitment or concern for others, interpersonal disharmony, nonreciprocated friendliness, and escape from or avoidance of intimacy.

    Each social relations measure was scored whenever the source mentioned that emotion in relation to the following:
    - his or her crewmates,
    - his or her own space agency,
    - the space agency in charge of the mission,
    - his or her own family, and
    - people in general.
3. **Coping strategies:** A standard set of coping categories was used to analyze the source materials.[36] These include both problem-oriented and emotion-oriented strategies. "Supernatural Protection" was added in our previous studies of Holocaust survivors and was retained in the current analysis.[37] This is not a coping strategy per se, but rather an expression of the individual's invocation of spirituality, religion, mysticism, or fatalism in dealing with problems (see table 3). The category was scored each time the narrative mentioned that the source had used that strategy in attempting to solve a problem.
4. **LIWC computer analysis:** In addition, material that was accessed through the Internet (oral histories and some interviews) was separately computer-analyzed by

---

34. J. R. McKay, "Affiliative Trust-Mistrust," in *Motivation and Personality: Handbook of Thematic Content Analysis*, ed. Smith, Atkinson, and McClelland, pp. 254–265.

35. D. P. McAdams, "Scoring Manual for the Intimacy Motive," *Psychological Documents*, vol. 2613 (San Rafael, CA: Select Press, 1984).

36. S. Folkman, R. S. Lazarus, C. Dunkel-Schetter, A. DeLongis, and R. Gruen, "Dynamics of a Stressful Encounter: Cognitive Appraisal, Coping, and Encounter Outcomes," *Journal of Personality and Social Psychology* 50 (1986): 992–1003.

37. P. Suedfeld, R. Krell, R. Wiebe, and G. D. Steel, "Coping Strategies in the Narratives of Holocaust Survivors," *Anxiety, Stress, & Coping* 10 (1997): 153–179.

Table 3. Coping categories and definitions.

	Coping Category	Definition
1.	Confrontation	Effort to resolve situation through assertive or aggressive interaction with another person
2.	Distancing	Effort to detach oneself emotionally from the situation
3.	Self-Control	Effort to regulate one's own feelings or actions
4.	Accept Responsibility	Acknowledging that one has a role in the problem
5.	Escape/Avoidance	Efforts to escape or avoid the problem physically
6.	Planful Problem-Solving	Deliberate (rational, cognitively oriented) effort to change or escape the situation
7.	Positive Reappraisal	Effort to see a positive meaning in the situation
8.	Seeking Social Support	Effort to obtain sympathy, help, information, or emotional support from another person or persons
9.	Endurance/Obedience/Effort	Trying to persevere, survive, submit, or comply with demands
10.	Compartmentalization	Encapsulating the problem psychologically so as to isolate it from other aspects of life
11.	Denial	Ignoring the problem, not believing in its reality
12.	Supernatural Protection	Invocation of religious or superstitious practices; efforts to gain such protection (e.g., prayer, amulets); reliance on luck, fate

Linguistic Inquiry and Word Count (LIWC).[38] LIWC is a word-count software application that identifies a variety of affective/emotional, cognitive, sensory/perceptual, and social processes, as well as references to personal space and orientation, motion, work, leisure, financial and metaphysical issues, and physical states. Because computer analysis is subject to many problems such as ignoring context and being restricted to those words and phrases that had been entered in the software dictionary, this was considered a secondary methodology in the current study.

---

38. J. W. Pennebaker, M. E. Francis, and R. J. Booth, *Linguistic Inquiry and Word Count (LIWC): LIWC 2001* (Mahwah, NJ: Erlbaum, 2001).

Results

We want to emphasize that the findings reported here concentrate on the impact of status as a member of the national majority or a minority within a space crew. Main effect differences as a function of minority-majority status based on characteristics other than nationality (e.g., gender, occupation, or job category on a space mission) are not reported.

In the findings described below, all cited differences were significant at $p = .05$ or better unless otherwise specified.

Value Hierarchies

1. **Preflight differences:** In references to their life before the mission, the 18 majority sources for whom we had complete data referred significantly more often to Achievement than did the 19 minorities.
2. **In-flight differences:** For the period of flight, internal analyses showed significant majority-minority differences on Achievement and Spirituality, with majority crewmembers higher on both.

Achievement and Conformity also showed interactions with nationality. Across all categories, Russians mentioned Achievement the most often. Americans ranked the highest on both values when flying in minority status, but Russians ranked highest when they were in the majority (see figure 1). With regard to Conformity, Americans were high when they were in the minority but low in the majority; Russians were the opposite (see figure 2).

Minorities and majorities also differed significantly as a function of flight duration. When discussing in-flight periods lasting over four months, minorities emphasized Security more than majorities: the reverse was true for short (less than two weeks) missions.

Of the 18 minority astronauts who made references to values while they were in space, 8 flew with predominantly American crews and 10 with Russian crews. Those who flew with the Americans showed significantly less Hedonism, Self-Direction, Conformity, and Security than those who flew with the Russians. Comparing minorities and majorities flying in American or Russian crews, we found an interaction: non-Americans flying with NASA mentioned Universalism more frequently than

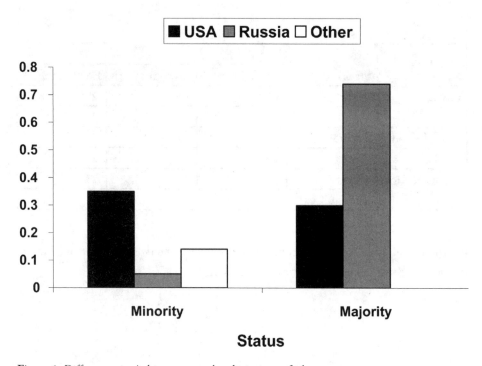

Figure 1. Differences in Achievement value during spaceflight.

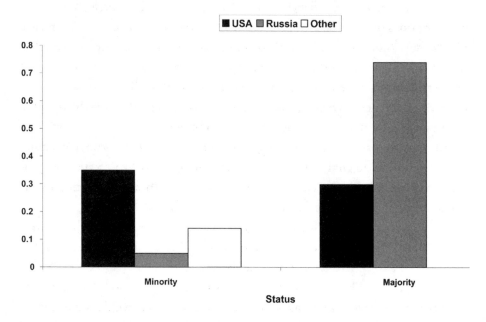

Figure 2. Differences in Conformity value during spaceflight.

Table 4. Phase- and status-related changes (mean value scores).

Value	Status	Mission Phase		
		Preflight	In-flight	Postflight
Power	Majority	.08	.06	.23
	Minority	.11	.05	.29
Achievement	Majority	.76	.47	.47
	Minority	.46	.21	.29
Self-Direction	Majority	.26	.17	.23
	Minority	.31	.10	.21

did their American colleagues; minorities flying with *Interkosmos* mentioned it less frequently than Russians in their own spacecraft.

Value Change

In general, all value references decrease when the source is discussing the period of his or her flight and mostly move back toward or above preflight baselines in descriptions of the postflight period. There were three significant phase-related value changes among the 17 minority and 15 majority crewmembers for whom we had complete data across the three mission phases (see table 4).

For minority sources only, significant changes over the three phases were found: increasing references to Spirituality and Hedonism between the in-flight and postflight phases, as well as decreasing references to Stimulation and Conformity over the three phases (see figure 3). For majorities, none of the changes was significant.

There were no significant interactions of gender and status or nationality and status with regard to value change over time, nor were there significant national differences in individual- versus collective-oriented values.

Social Relations

There were four separate, although correlated, measures within this category: Trust, Mistrust, Positive Intimacy, and Negative Intimacy.

Flying with Strangers: Postmission Reflections of Multinational Space Crews

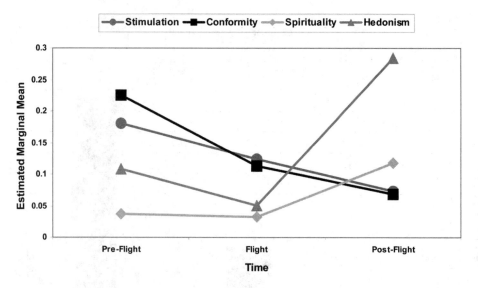

Figure 3. Minority value changes by mission phase.

1. **Majorities versus minorities:** Minority or majority status made a significant difference in the social relations references: minorities were more mistrustful and more negative about intimacy than majorities. There were no reliable overall differences and no interaction effects on the basis of gender or the nationality of crewmates. References to relations between the source and his or her family members were more positive in both subcategories when the source flew with a foreign majority. Minority astronauts showed more Negative Intimacy references toward their own space agency and showed more Mistrust toward the foreign home agency of their majority colleagues.

   There were no majority-minority differences in either positive or negative orientation toward astronauts' fellow crewmembers; these comments were predominantly positive (high Trust and Positive Intimacy). The same was true of references to other people in general. Minorities who flew in Russian spacecraft mentioned both positive and negative social relations (Trust and Mistrust, Positive and Negative Intimacy) more frequently than those who flew with American majorities, although only the difference in Trust reached the $p = .05$ level of significance.

2. **Mission duration:** Regardless of status, long-duration crewmembers made more references to Mistrust than those on shorter missions. Majorities had higher

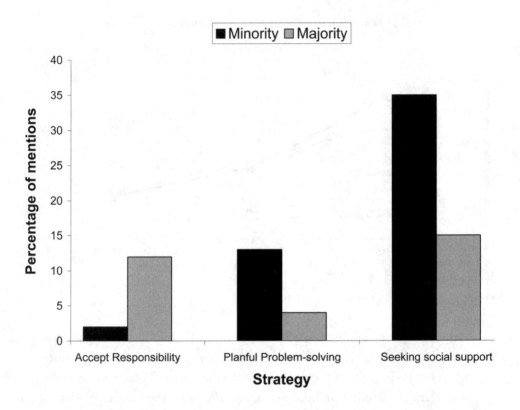

Figure 4. Coping strategies by minority-majority status.

Mistrust and higher Negative Intimacy scores than minorities when discussing short missions, but minorities on long missions were more negative than their hosts.

3. **Mission phase:** Disaggregating minority and majority crewmembers, we found that both decreased in Trust from the pre- to the postflight portion of their narrative, but the drop was marginally ($p = .08$) steeper among minorities. On Mistrust, majorities remained stable while minorities showed a dramatic increase, especially from the in-flight to the postflight stages. There was also a marginal ($p = .07$) interaction effect on Negative Intimacy, with minorities showing a steep increase (again, especially from in- to postflight), while majorities started out much higher in the preflight stage and remained stable at that level.

Coping Strategies

Coping data were collected from 56 astronauts and cosmonauts. There were no significant overall differences for own nationality, gender, or nationality of crew colleagues, and there were no significant baseline (preflight) differences. Over all mission phases and nationalities, there were three statistically significant majority-minority differences (see figure 4).

The remainder of the coping strategy analyses concentrated on descriptions of the mission phase where coping was most crucial—that is, during the flight. There was a significant majority-minority main effect for 3 of the 12 categories. Majority crewmembers were higher than minorities on Accepting Responsibility and lower on Planful Problem-Solving and Seeking Social Support.

Mission duration had significant effects on five coping strategies during flight, in each case with long-duration (four months or more) fliers higher than the short (two weeks or less). The strategies affected were Confrontation, Escape/Avoidance, Denial (all $p = .01$), Accepting Responsibility ($p = .05$), and Supernatural Protection ($p = .06$).

Still during the flight phase, duration also figured in four interactions with majority-minority status. In each case, the difference appeared in the long-duration group only: Accepting Responsibility, Denial, and Escape/Avoidance ($p = .07$), with majority fliers the highest on all three measures, and Supernatural Protection ($p = .06$), with minority crewmembers the highest.

LIWC

Analysis by nationality showed higher word count scores for Russians than Americans on references to affect in general, positive emotions, and optimism. Americans scored higher on references to social interaction. Word count differences as a function of majority-minority status showed that minorities used fewer words and phrases referring to social interaction, community, other individuals, and human beings as a group.

There were no significant differences on such LIWC categories as anxiety, anger, or sadness and no differences as a function of gender or mission duration.

DISCUSSION
Flying with Strangers: The Influence of Minority-Majority Status

It seems clear that space voyagers who fly in a crew composed mostly of people from their own country have a different experience from those who are a minority flying with a mostly foreign crew. However, contrary to some assumptions that minority status would be generally aversive, the data show a mixed picture. For example, among minority participants, the value of Stimulation and Conformity decreases between the preflight and in-flight phases; this decrease presumably indicates that both boredom and the desire to submerge one's own culturally learned characteristics become less of a challenge over time. Simultaneously, Spirituality increases, indicating a growing internal recognition of transcendental values that is often found among astronauts and is apparently not thwarted—and may in fact be enhanced—by being the "odd person out" in the crew.[39] Hedonism also increases, implying a heightened concern with pleasure.

Being in the minority was associated with fewer references to social interaction, community, other specific individuals, and human beings in general. This datum emerged from the computerized frequency analysis and is difficult to interpret because LIWC merely counts words and phrases; it does not differentiate on the basis of context. More interesting is the fact that minority status also led to more positive comments about one's family, perhaps to compensate for some degree of social isolation; apparently, absence made the heart grow fonder (which was not found for majority crewmembers). One's own home organization evoked more negative references, confirming the complaints of inadequate preparation and support that characterize some anecdotal comments. The agency in charge of the mission—that is, a space agency foreign to the minority flier—was viewed with increasing mistrust as the mission unfolded, perhaps with the recognition that its rules and procedures were alien and sometimes uncomfortable.

However, there was no evidence that bad feelings prevailed toward the majority crewmates, again despite conclusions sometimes drawn from selected anecdotal reports. In fact, the data showed a generally trustful and friendly attitude,

---

39. P. Suedfeld, "Space Memoirs: Value Hierarchies Before and After Missions—A Pilot Study," *Acta Astronautica* 58 (2006): 583–586.

compatible with such reports as that of Shannon Lucid, a long-duration *Mir* resident with two Russian crewmates: "Yes, we really had a good time together. We really enjoyed being there together. Yuri and Yuri were absolutely fantastic to work with. I mean, I could not have picked better people to spend a long period of time with. We just lived every day as it came. We enjoyed every day. We enjoyed working together and joking around together. It was just a very good experience, I think, for all of us."[40]

It should be noted that, in the same way, the majority crewmembers expressed trust and friendship toward their foreign colleagues—once again contradicting the negative picture drawn from selective quoting of particular complaints. However, comparisons of comments concerning the in-flight phase with those concerning the postflight phase showed that these positive feelings did decline on both sides (and especially among minorities), and both majority and minority veterans of long-duration missions showed more Mistrust and Negative Intimacy than those who flew shorter missions. Growing interpersonal stress as a function of isolation and confinement with the same small group for over four months was thus confirmed by our data.

We expected to find changes in values, as in previous research.[41] Among the most interesting changes was the drop in references to Power and Self-Direction, for both groups but especially for the minority, as the narratives moved from the pre- to the in-flight portion, followed by increases after the flight. The highly regimented aspects of the launch and the flight itself probably explain the general finding, and the somewhat tenuous and isolated role we have seen for many minority crewmembers, from which they were freed after the mission, explains their more dramatic changes. Minorities' position within the crew may also be implied by their higher scores on coping by Seeking Social Support and lower scores on Accepting Responsibility—which in many cases they were not permitted to do. However, they also used Planful Problem-Solving more frequently than their majority counterparts,

---

40. Shannon Lucid, interview by Mark Davison, Rebecca Wright, and Paul Rollins, 17 June 1998, published through "Shuttle-Mir Oral History Project," Johnson Space Center History Portal, available at *http://www.jsc.nasa.gov/history/oral_histories/participants.htm* (accessed 7 June 2010).

41. Suedfeld, "Space Memoirs": 583–586.

perhaps because they had to face not only the problems of spaceflight, but also the problem of gaining full social equality.

Majority crewmembers' characterization of their pre-mission life was marked by more references to Achievement than that of the minority members; striving to become an astronaut may be a more vivid achievement goal for those hopefuls whose country has its own spaceflight capability (all majority subjects were from either the United States or the USSR/Russia). Achievement scores in general were high, compared to those for other values, as one would expect from a group with the high levels of achievement that spacefarers had reached even before becoming astronauts or cosmonauts.[42]

We speculate that for most astronauts and cosmonauts, Achievement is a high-level background variable that tends to be taken for granted, not a primary concern, except at particular periods; preparing to become an astronaut and then to embark on a space mission may be such periods. This hypothesis is supported by the finding of overall decreases in Achievement references between the preflight phase and both subsequent phases as the successful mission and return reduce concern about the person's ability to function at the desired level. In fact, space agencies may want to provide Achievement opportunities for postflight astronauts to help them regain their interest in this value, as an astronaut for whom Achievement has become drastically less important than before may suffer serious adverse consequences in adjustment, health, and performance.[43]

Status and Nationality

The often-cited cultural differences—especially between the two major spacefaring nations, the United States and the USSR/Russia—seemed to make no difference as to how positive and trusting relations among crewmembers were. Neither were they reflected in overall comparisons of the individual-oriented versus group-oriented values. However, Russian cosmonauts were higher in mentions of Achievement than were American or other astronauts.

---

42. P. Suedfeld and T. Weiszbeck, "The Impact of Outer Space on Inner Space," *Aviation, Space, and Environmental Medicine* 75, no. 7, supplement (2004): C6–C9.
43. Suedfeld and Weiszbeck, "The Impact of Outer Space on Inner Space": C6–C9.

In some of our data, nationality did interact with the majority or minority status of the crewmember. For example, it is intriguing to see Russians mentioning Achievement-related values especially often when they were in the majority during a space mission, but much less frequently when they were in the minority, whereas Americans did not change much as a function of their status. Achievement for the cosmonauts seems to be more closely linked to social approval from their compatriots than it is for astronauts, perhaps stemming from the collectivist-individualist difference.

Astronauts and cosmonauts varied in references to Conformity, but in opposite directions: the former were slightly higher when they were in the minority, and the latter were much higher when they were in the majority. It may be that Americans felt somewhat easier about being different from their compatriots but felt constrained to fit into their foreign crewmates' expectations. Russians, to the contrary, confirmed traditional mores when they formed the majority but were freer with dissent or individualism when flying alone with foreigners.

Status and Flight Duration

Discussing the spaceflight experience itself, majorities on short-duration missions referred to issues of Security more than minorities, perhaps because they were the hosts responsible for the safety of the capsule and the mission, and they may have felt responsible for the welfare of their guests as well as themselves. However, as missions lengthened, the pattern was reversed; the hosts may have come to feel more secure while the guests became more concerned, possibly because of problems that only the former knew how to solve (e.g., the gradual mechanical deterioration of *Mir*) or possibly because the latter were being prevented from full engagement in meaningful work dealing with those problems.

References to coping strategies in narratives of short flights showed no differences as a function of status. Long missions, however, evoked a majority-minority difference, with the majority higher on mentions of Accepting Responsibility, Denial, and Escape/Avoidance. It may be that these went together: as one accepted more responsibility for solving problems, it may also have become more important to use emotion-oriented means of coping to reduce psychological stress. Minorities mentioned Supernatural Protection more frequently, confirming their increased mention of spiritual values.

Minorities participating in short flights may have appreciated the experience so much that they let personal disharmony pass without comment, while the majority felt less compunction about mentioning interpersonal problems with the salient "other." In long flights, the guest may have both experienced more abrasiveness and become less reluctant to describe it later; the hosts may have become habituated to the strangeness of the visitor or (note the increased mentions of Escape/Avoidance and Denial, mentioned above) withdrawn from unpleasant interactions either physically or psychologically, or both.

Status and Host Nationality

Minority members also reacted differently depending on which nation constituted the majority. Minorities who flew in predominantly American crews cited pleasure and enjoyment, security, autonomy, and (paradoxically) conformity less frequently than those who flew with Russians and also made fewer valenced (positive and negative) references in either direction to social relationships. Whether these differences were the result of the minority member's changing to fit in with the majority (implying that the Russian colleagues themselves were more expressive than American crews, as is also indicated by the LIWC results) or of asserting their own cultural distinctiveness is impossible to tell.

Minorities who flew with *Interkosmos* expressed a feeling of global concern for Earth (universalism) less frequently than their Russian hosts, while the reverse was true for minorities who flew with NASA. However, this is a misleading datum: the two groups of minority fliers did not differ from each other; it was the hosts who differed, with higher scores among the Russian than the American majorities. Whether this is a function of the generally greater emotional expressivity of the Russians or is specific to the topic is not revealed by our data.

CONCLUSION

It appears that any problems related to mixed-nationality space crews may be more a function of the fact that space capsules have belonged to, and were predominantly operated according to the traditions and standard operating procedures of, one space agency. As a result, the minority "guest" tends to feel left out, unfamiliar with important matters that come naturally to his crewmates, and also feels neglected and let down by his own home organization. This may be an even more pressing problem for minorities who can fly only in that status—that is, all nationalities except Russians, Americans, and (perhaps soon) Chinese. The development and use of truly international missions, including international vehicles and common procedures, is a necessary countermeasure.

One possible way to reduce misunderstanding, miscommunication, and cultural friction would be for all mission participants, both space crews and ground staff, to have in situ language training and familiarization in each other's countries. When astronauts were being prepared for the Shuttle-*Mir* missions, they underwent extensive Russian language training and spent considerable time in Russia, both training and socializing with their future crewmates. Apparently it was not considered necessary for the cosmonauts in the crew to have equal exposure to American culture and folkways: cosmonauts did not have prolonged deployments to Houston to become linguistically and culturally adapted. This omission may have been economical in terms of money and time, but it was shortsighted in terms of smoothing performance and interpersonal relations in space, and the lack of similar provisions for mission controllers and staff exacerbated the problems.

We did not look at cultural or personal differences based on characteristics other than nationality, although they have also been thought of as causes of increased stress. However, there is no a priori reason to suspect that they would be any more important in that role than nationality itself, given its pervasive nature: it underlies language, values, history, traditions, child-rearing approaches, political ideologies, concepts of human nature and the individual-society relationship, and so on. We have found, as have other researchers, that differences within each national group are greater than differences across groups; but the latter differences in any case were few except as they interacted with majority-minority (or host-guest) status.

As has been expected, based on anecdotes but without much empirical grounding, long-duration missions (four months or more) reveal more abrasiveness and dissatisfaction.[44] Our data show that these negative tendencies also include more

---

44. See, for example, Lebedev, *Diary of a Cosmonaut*, or Lebedev, cited in Zimmerman, *Leaving Earth*, p. 134.

divergence in values and concerns, and on the part of the majority, emotion-oriented coping that does not really solve problems effectively. How this would develop further on voyages lasting several years is unknown, but is certainly something that space agencies need to think about.

Individual changes in astronaut personality—in values and social orientation—may be evanescent, persistent, or permanent. They may be particularly problematic for people who fly as minorities, especially on long missions. If they last into the postflight life of the crewmember, they may affect his or her family relationships, career progress, and physical and mental health. Again, it behooves the home organization to conduct nonthreatening and supportive post-return help where needed, both for astronauts and for their families.

POSTSCRIPT

Much of the research concerning international crews has been based on the prospect that such crews will continue to be the norm, as they have been on the International Space Station. Some commentators have asserted that a project as massive and complex as the trip to Mars would have to be an international technical, scientific, and financial effort (although that does not necessarily imply a multinational crew). Politically, it seems that cooperation and collaboration have become the permanent hallmark of space exploration.

As this chapter was being written, the old space race was showing signs of reviving. NASA's three-stage plans, sparked by President George W. Bush, had turned the world's space agencies in new directions. Human return to the Moon, a Moon base, and the voyage to Mars, seemed to have been adopted as goals by all of the major space agencies. But by the same token, several agencies announced a hope that *their* astronauts would be among those who took these giant steps. Some countries (e.g., Canada) accepted that this would happen on an international vehicle, but others (e.g., Russia and China) indicated plans to go on their own. At that time, the Administrator of NASA deplored the possibility of Chinese "taikonauts" reaching the Moon before Americans return to it—an echo of the early years, when competition was the name of the game.

The Obama administration's 2010 decisions concerning the near future of NASA's human space exploration program—canceling the construction of new space vehicles, delaying if not abandoning a return to the Moon, delaying a voyage to Mars, and introducing the novel possibility of using an asteroid as the next new destination—may have put the United States on the sidelines in any such renewed space race. The impact of these changes for the future of multinational space crews remains to be seen.

Chapter 8
Spaceflight and Cross-Cultural Psychology

Juris G. Draguns
  Department of Psychology
  Pennsylvania State University

Albert A. Harrison
  Department of Psychology
  University of California, Davis

ABSTRACT

In the first decade of spaceflight, the United States and the Soviet Union were locked in relentless competition, but in 1975, the two nations joined together for the Apollo-Soyuz Test Project. In 1978, the Soviets began their *Interkosmos*, or "guest cosmonaut," program, whereby non-Soviet cosmonauts, mostly drawn from Eastern Bloc nations, joined Soviet crews on Salyut space stations. Meanwhile, in 1969, the United States invited Europeans to participate in post-Apollo flights, and the Europeans developed Spacelab, which first flew with the Space Shuttle *Columbia* in 1983. Over the years, the largely symbolic *Interkosmos* program grew into flights involving true partnerships between Soviets and non-Soviets, while U.S. flights drew payload specialists from many different lands. In the early 1990s, astronauts joined cosmonauts on *Mir*, and today the International Space Station routinely carries multicultural crews. Experience gained during early international missions revealed problems in such diverse areas as mission organization and management, work habits, communication, interpersonal relations, privacy, personal cleanliness habits, food preferences, and leisure-time activities. We introduce the culture assimilator as a potential aid in preparing spacefarers for international missions. We then explore cultural dimensions based on worldwide studies of values in work environments and trace their implications for international flights. To conclude, we sound a note of caution against reifying cultural differences lest they give rise to harmful stereotypes. Even as international missions will benefit from cross-cultural psychology, cross-cultural psychology will benefit from studies conducted in space.

SPACE EXPLORATION AND CULTURALLY SHAPED BEHAVIOR: ANTICIPATIONS AND PREPARATIONS

Both spaceflight and cross-cultural psychology are young endeavors. As noted in the introductory chapters, more than a few test pilots, high-altitude balloonists, and rockets (including some carrying test animals as passengers) preceded Yuri Gagarin and Alan Shepard into space in 1961. Like spaceflight, systematic and continuous pursuit of research in cross-cultural psychology has a lengthy set of scattered antecedents.[1] The inaugural meeting of the International Association for Cross-Cultural Psychology took place in Hong Kong in 1972.[2]

The concerns of these two ventures began to coalesce in 1975, when the Cold War defrosted a bit and the United States and Soviet Russia worked together on the Apollo-Soyuz Test Project. This required substantial cooperation among project managers and engineers to ensure compatibility of the U.S. and Soviet spacecraft; it also required collaborative training of astronauts and cosmonauts. Participants grappled with each other's languages and customs, as well as each other's technology. Occasionally, astronauts and cosmonauts would slip away from program officials for hunting trips and other enjoyable, morale-building activities.[3] As NASA historians point out, "The flight was more a symbol of the lessening of tensions between the two superpowers than a significant scientific endeavor, a sharp contrast with the competition for international prestige that had fueled much of the space activities of both nations since the late 1950s."[4]

---

1. G. Jahoda, "Our Forgotten Ancestors," in *Nebraska Symposium on Motivation*, vol. 37, *Cultural Perspectives*, ed. J. J. Berman (Lincoln: University of Nebraska Press, 1990), pp. 1–40.
2. J. L. M. Dawson and W. J. Lonner, eds., *Readings in Cross-Cultural Psychology* (Hong Kong: University of Hong Kong Press, 1974).
3. D. K. Slayton and M. Casutt, *Deke! An Autobiography* (New York: Forge Books, 1995).
4. NASA, "Chronology of Selected Highlights in the First 100 American Spaceflights, 1961–1995," *http://history.nasa.gov/Timeline/100flt.html* (accessed 7 April 2008).

The initially daring and innovative enterprise of international missions has expanded greatly over the years, although approximately 20 years would pass before Americans and Russians again flew together. However, international crews appeared early on. In 1976, the Soviet Union announced its *Interkosmos* or "guest cosmonaut" program. As James Oberg explains, the first, three-week flights in March, June, and August 1978 included Czech Air Force Pilot Vladimir Remek, Polish pilot Miroslaw Hewrmaszewski, and East German Air Force officer Sigmund Jahn.[5] Actual flying was done by Soviet cosmonauts, who had many more years of training than their guests. Oberg recounts a joke about Remek returning to Earth with "red hands" disease. When flight surgeons asked how he had acquired this malady, Remek explained, "Well, in space, whenever I reached for this or that switch, the Russians cried 'Don't touch that!' and slapped me on my hands."[6] Other guest cosmonauts were recruited from Cuba, Bulgaria, Mongolia, and Romania.

Although Oberg portrays *Interkosmos* as largely symbolic, he also predicted—correctly—that this could give rise to true international missions with more active participation on the part of the international partners. More recently, a NASA report pointed out that Russia has flown cosmonauts from many countries, including Germany, India, Italy, Japan, South Africa, Syria, the United Kingdom, the United States, and Vietnam.[7] The first Western European to fly, Frenchman Jean-Loup Chrétien, did so aboard Russian craft in 1982.

In the late 1970s, while the *Interkosmos* program flourished, the last Apollo rocket had been launched, but the Space Shuttle had not yet completed the transition from drawing board to orbit. In 1969, NASA had invited Europeans to participate in post-Apollo flights, and in the 1970s, while the Europeans were hard at work designing Spacelab, Ulf Merbold from West Germany, Wubbo Oeckels from Holland, and Claude Nicollier from Switzerland were training for the Shuttle. Merbold, flying aboard *Columbia*, was the first to reach orbit and did so in November 1983. By 2003, the 25th anniversary of Remek's historic flight, 30 European astronauts had participated in 44 missions: 26 in collaboration with NASA and 18

---

5. J. Oberg, "Russia's 'Guest Cosmonaut' Program, A Commentary," *L-5 News* 3, no. 11 (November 1978): 1–2.

6. Ibid., p. 2.

7. "Astronaut," *World Book at NASA*, http://www.nasa.gov/worldbook/astronaut_worldbook.html (accessed 23 May 2010).

aboard Russian spacecraft.[8] As of early 2008, Canadian astronauts had participated in 1 Soyuz and 13 Space Shuttle flights.[9] Five Japanese astronauts had successfully completed their missions, and on 9 April of that year, the first South Korean astronaut, Yi So-Yeon, blasted off with two cosmonauts to head to the ISS.[10] With 29 partners supporting the International Space Station, multicultural crews are the rule, rather than the exception.

International flights make tremendous sense. First, as Jennifer Boyd Ritsher points out, by drawing from an international pool, managers can tap a broader range of interests and skills.[11] Second, space missions are, in effect, overarching or superordinate goals that encourage different nations to work together and may serve as a prototype for other collaborative ventures. And, most obviously, international cooperation defrays the enormous expense, increasing the palatability of flight for large, wealthy nations and enabling nations with fewer resources to participate in space. Differences based on ethnicity and nationality have implications for everything from international relations to tourism. It is not surprising, therefore, to speculate that cultural differences could affect safety, performance, and well-being in space. Since the United States and Russia control access to space, relations between U.S. astronauts and Russian cosmonauts have gained the most attention.[12]

We found little discussion of cultural factors in early space psychology papers, perhaps because during the tensions of the 1960s, it was all but impossible to imagine astronaut-cosmonaut collaboration. By the mid-1980s, in recognition of the Apollo-Soyuz Test Project, *Interkosmos*, and Spacelab, cultural factors had joined the list of psychologists' concerns, with the threat of miscommunication (both verbal and nonverbal) the most prominent worry.[13] Yet already by that time, there were

---

8. J. Feustl-Beuchl, "25 Years of European Human Spaceflight," *ESA Bulletin* (November 2003): 6–16.

9. "Canadian Astronaut Missions," http://www.asc-csa.gc.ca/eng/astronauts/missions.asp (accessed 23 May 2010).

10. JAXA, "JAXA's Astronauts," *Human Space Activities*, http://www.jaxa.jp/projects/iss_human/astro/index_e.html (accessed 6 April 2008).

11. J. B. Ritsher, "Cultural Factors and the International Space Station," *Aviation, Space, and Environmental Medicine* 76, no. 6, sect. II (2005): B135–B144.

12. Ibid.

13. M. M. Connors, A. A. Harrison, and F. R. Akins, *Living Aloft: Human Requirements for Extended Spaceflight* (Washington, DC: NASA SP-483, 1985).

other concerns as well—prejudices against members of other cultures, in addition to conflicting values and preferences. How would emotionally controlled astronauts react to highly expressive Russians?[14] Interest was further piqued in the 1990s when astronauts joined cosmonauts on *Mir*. A 1993 study based on debriefing American astronauts who had flown with international crewmembers revealed 9 preflight, 26 in-flight, and 7 postflight incidents of misunderstanding, miscommunication, and interpersonal conflict.[15] Journalist Bryan Burrough described (and perhaps slightly sensationalized) several instances where cultural differences influenced astronaut performance and morale.[16] Cultural problems that the astronauts reported pertained to personal hygiene, food preferences, and chosen activities as well as to interpersonal distance, privacy, and work styles.[17] Jason Kring summarized experience with international crews prior to the ISS as follows:

> Apollo-Soyuz, Shuttle, and Shuttle-Mir proved, for the most part, that international cooperation was feasible and rewarding. During Shuttle-Mir, the U.S. and Russia gained from each other's strengths and weaknesses, learned how to work together, and gathered insights that directly benefitted the ISS. There were, however, moments of confusion and disagreement between the two countries, incidents that affected crew performance and mission activities. Differences in management, training, decision-making and problem resolution, for example, were tied to differences in national culture and the backgrounds of the U.S. and Russian crewmembers and personnel.[18]

A recent International Academy of Astronautics study group pointed out that one of the reasons that culture is so important is because it influences almost all

---

14. Ibid.

15. P. A. Santy, A. W. Holland, L. Looper, and R. Macondes-North, "Multicultural Factors in an International Crew Debrief," *Aviation, Space, and Environmental Medicine* 64 (1993): 196–200.

16. B. Burrough, *Dragonfly: NASA and the Crisis Aboard Mir* (New York: Harper Collins, 1998).

17. Ritsher, "Cultural Factors and the International Space Station."

18. J. Kring, "Multicultural Factors for International Spaceflight," *Journal of Human Performance in Extreme Environments* 5, no. 3 (2001): 11.

forms of behavior.[19] For example, in addition to causing communications difficulties, culture influences psychological reactions to stress, shapes social behavior, and molds attitudes in such areas as food preferences, recreational activities, humor, and privacy. Culture affects the likelihood and kinds of mental illness that arise, as well as diagnosis and treatment. A severe problem that would be instantly recognized by a psychiatrist who shared the same cultural background as the person in crisis might not be recognized by a therapist from a different culture. Similarly, forms of therapy that work in one culture may be ineffective in another. In his comprehensive summary, Kring identified 10 areas related to spaceflight that are influenced by national culture and background: "(a) communication; (b) cognition and decision making; (c) technology interfacing; (d) interpersonal interaction; (e) work, management, and leadership style; (f) personal hygiene and clothing; (g) food preparation and meals; (h) religion and holidays; (i) recreation; and (j) habitat aesthetics."[20] A 2007 review of training for astronauts identified cultural awareness and sensitivity as among the skills required for astronauts.[21] One of the tools that it mentions, in passing, is the culture assimilator, to which we shall turn shortly.

As early as the 1970s, then, NASA and its Soviet, later Russian, counterpart agency were faced with the major challenge of ensuring that astronauts and cosmonauts would work together smoothly and effectively. To this end, it was imperative to identify and overcome obstacles rooted in socialization within different cultures and contrasting sociopolitical systems. They had to recognize and accommodate the weight of lifelong training and indoctrination, inextricably tied to the divergent traditions and ideologies of the two countries. Today, prior to an international mission, astronauts receive at least a one-day seminar that discusses customs and cross-

---

19. N. Kanas, G. Sandal, J. B. Ritsher, V. I. Gushin, D. Manzey, R. North, G. Leon, P. Suedfeld, S. Bishop, E. R. Fiedler, N. Inoue, B. Johannes, D. J. Kealry, N. Kraft, I. Matsuzaki, D. Musson, L. A. Palinkas, V. P. Salnitskiy, W. Sipes, and J. Stuster, "Psychology and Culture During Long Duration Space Missions" (revised final report, International Academy of Astronautics Study Group on Psychology and Culture During Long-Duration Missions, 28 November 2006).
20. Kring, "Multicultural Factors for International Spaceflight," p. 11.
21. S. J. Hysong, L. Galarza, and A. W. Holland, "A Review of Training Methods and Instructional Techniques: Implications for Behavioral Skills Training in U.S. Astronauts" (NASA TP-2007-213726, May 2007), available at *http://ston.jsc.nasa.gov/collections/TRS/_techrep/TP-2007-213726.pdf* (accessed 2 July 2010).

cultural issues pertaining to the nation of the international partner. Additionally, there is an opportunity for astronauts in Long Duration Mission (LDM) training for the ISS to live with a Russian family for a few weeks during their years of ISS training. Many astronauts have taken advantage of this experiential opportunity. Also, the Behavioral Health and Performance Section at Johnson Space Center developed a two-day seminar for LDM astronauts. Developed and presented with astronaut participation, this program is now managed by the Astronaut Corps.[22] But perhaps most important of all is the joint training, for many years, of astronauts and cosmonauts in the United States and Russia.

When initially confronting cultural issues, NASA had to rely on intuition, trial and error, and common sense. Now, NASA can draw on the results of several decades of research on interaction in small groups, especially in those of heterogeneous composition, to better understand the social realities within the space capsule. Researchers and operational personnel now build on the store of information that accumulated in preparing sojourners, visitors, and immigrants for encounters with a new culture. The space program can both utilize and anticipate the achievements of emerging academic cross-cultural psychology, which is based in part on controlled observation.[23] Thus, we see a shrinking gulf between the research of cross-cultural psychologists and NASA's efforts to prepare astronauts for international missions. We turn now to theories and principles that may reduce this gulf further.

THE DOMAIN OF CROSS-CULTURAL PSYCHOLOGY: CULTURAL CHARACTERISTICS AND INTERCULTURAL INTERACTION

For the present purposes, following Kring, culture is defined as "an individual's values, beliefs, behavior patterns and language that are directly linked to his or her national and ethnic background."[24] Cross-cultural psychology is concerned with "the study of similarities and differences in individual psychological functioning

---

22. W. Sipes, personal communication, 8 April 2008.
23. Ritsher, "Cultural Factors and the International Space Station."
24. Kring, "Multicultural Factors": 12.

in various cultural and ethnic groups; of the relationships between these variables and socio-cultural, ecological, and biological variables, and of ongoing changes in these variables."[25] In addition to pursuing these objectives, some psychologists have ventured into intercultural research, which is concerned with the study of contacts and interactions between persons from different cultures.[26] Intercultural encounters can be better understood and their effectiveness can be enhanced on the basis of the body of knowledge of cross-cultural psychology that has largely accumulated in the course of the last three or four decades. We now turn to some of the approaches that have evolved from this research.

The Culture Assimilator

The culture assimilator, developed by F. E. Fiedler, T. Mitchell, and H. C. Triandis, is an extensively used and systematically validated programmed learning approach to cross-cultural training.[27] As described by Triandis, "it consists of 100 to 200 episodes, i.e. scenarios where people from two cultures interact. Each episode is followed by four or five explanations of why the member of the other culture has acted in a specific way. The trainee selects one of the explanations and is asked to turn to another page (or computer screen) where feedback is provided concerning the chosen explanation."[28] The objective is not only to help people anticipate concrete situations that are construed and reacted to differently in another culture, but to help them gradually grasp the rules that underlie the various expected and "normal" courses of action in the culture in question. Controlled studies of training by means of the culture assimilator demonstrate that it results in trainees' selecting explanations of others' behavior that are closer to those chosen by the members of the target culture. Moreover, the learners' attributions tend to become more spe-

---

25. J. W. Berry, Y. H. Poortinga, M. H. Segall, and P. R. Dasen, *Cross-Cultural Psychology: Research and Applications*, 2nd ed. (New York: Cambridge University Press, 2002), p. 3.

26. R. W. Brislin, *Understanding Culture's Influence on Behavior*, 2nd ed. (Fort Worth, TX: Harcourt College Publishers, 2000).

27. F. E. Fiedler, T. Mitchell, and H. C. Triandis, "The Culture Assimilator: An Approach to Cross-Cultural Training," *Journal of Applied Psychology* 55 (1971): 95–102.

28. H. C. Triandis, *Culture and Social Behavior* (New York: McGraw-Hill, 1994), p. 278.

cific and complex and less ethnocentric, and they become better at estimating what members of the target culture will do.[29] Thus, a great deal of anticipatory culture learning takes place in a relatively short period of time.

The culture assimilator then represents a vicarious enculturation experience preparatory to the actual encounter with a new and different culture. Does it have anything to offer over the real-life immersion in the other culture, followed by thorough debriefing that NASA has already instituted for astronauts assigned to multinational space crews? Potentially, there are four advantages that the assimilator may provide.

First, the rationale, if not the actual procedure, of the culture assimilator pinpoints and makes explicit the themes and features of the other culture. This is particularly germane to the debriefing phase as specific incidents and encounters come to be imbued with more general meaning and patterns are established.

Second, properly constructed, the culture assimilator increases the range of situations that people are likely to encounter. This helps people "fill in the gaps" and respond appropriately to spontaneous and somewhat unpredictable instances of unfamiliar cultural realities.

Third, on the basis of the accumulated experience with mixed American-Russian space crews over three decades, we can anticipate specific problems and incorporate them into custom-made versions of the assimilator. That is, the assimilator can be adapted to cover the unique situations of astronauts and cosmonauts.

Fourth, the culture assimilator methodology is useful in both its specific and its general aspects. Given the ever-greater variety of nations from which spaceflight participants are increasingly drawn, it may be possible to develop a general culture for increasing generic awareness and sensitivity, especially into the early phases of astronaut training.

Subjective Culture

Conceptually, the development of the culture assimilator was linked with basic research in Greece, India, and Japan, which Triandis conducted on subjective

---

29. Ibid.

culture.[30] Subjective culture encompasses the characteristic ways of viewing the humanmade environment.[31] More specifically, it comprises the standards by which events and actions are evaluated in a culture as well as the ideas, theories, and explanations shared by that culture. The investigation of subjective culture focuses on categorizations, associations, and beliefs, and on their interrelationships. In this manner, it becomes possible to ascertain the inner coherence of culture as it is experienced by its members. Subjective culture is implicitly known to its members, even though this knowledge is rarely articulated. Somewhat like a speaker's first language, it is used effectively and correctly by the members of the culture, but its rules and practices must be made explicit for an outsider.

Triandis developed a number of complex methods for bringing facets of subjective culture to the fore.[32] Thus, beliefs, behavior patterns, and social rituals are related to the other features of the culture. In this manner, the culturally characteristic meanings of "happiness," "risk," "science," "family," and other terms and concepts are established empirically and related in pinpoint fashion to behaviors, practices, and convictions.

In reference to spaceflight, it is interesting to explore the commonalities and differences in association, categorization, and rules of conduct among the astronauts of various nations. Specifically, it would be worthwhile to investigate cross-culturally the meaning of "spaceflight." Into what clusters of meaning is it embedded, and what are the behavioral consequences? What cultural misunderstandings, if any, may arise on the basis of putative cultural divergences related to these notions? Conversely, is it possible to think differently but act identically, especially when the participants from several cultures are thoroughly trained in their technical tasks and their interactions?

Subjective culture allows investigators to peer beneath the behavioral surface and to uncover the implicit, yet socially shared, levels of consciousness in their relation to characteristic behavior. Just as no two persons speak their first language alike or perfectly, each person's subjective culture is the best approximation of his or her culture based on the individual's lifetime of experience. To optimize the intense and demanding interaction in the space capsule, we can ascertain the subjective

---

30. H. C. Triandis, *Subjective Culture* (New York: Wiley, 1972).
31. Ibid.
32. Triandis, *Culture and Social Behavior*.

cultures of specific individuals in their personal uniqueness, cultural distinctiveness, and human universality.

Critical Incidents

K. Cushner and R. W. Brislin extended culture assimilator methodology to a set of 100 sketches pertaining to 18 themes representative of various domains of intercultural experience.[33] These vignettes represent critical incidents frequently encountered by newcomers to a culture other than their own. Each vignette is accompanied by four options, one of which, through thorough pretesting, has been deemed to be the most culturally appropriate. Critical incidents serve as points of departure for the development of more specific, yet critically important, items that would sensitize astronauts to crucial, yet culturally divergent, aspects of interaction and cooperation during space missions. Extending these methods to space, we would pool knowledge of participants' cultures with information on the personal, social, and technical challenges of flight. That is, previously encountered problems, conflicts, and crises are the requisite "raw material" for the construction of incidents.

Cultural Dimensions: Value Orientation

On the basis of multistage factor analyses conducted on a huge store of data from 72 countries, G. Hofstede identified four relatively independent cultural dimensions.[34] Originally based on questionnaire responses on values in the workplace, Hofstede's four factors have been studied around the world in educational, social, mental health, and many other settings. The second edition of his monograph and more recent publications, for example by G. Hofstede and G. J. Hofstede,

---

33. K. Cushner and R. W. Brislin, *Intercultural Interactions: A Practical Guide*, 2nd ed. (Thousand Oaks, CA: Sage Publications, 1995).

34. G. Hofstede, *Culture's Consequences: International Differences in Work-Related Values* (Beverly Hills, CA: Sage Publications, 1980).

document a prodigious amount of research-based findings.[35] It is no exaggeration to say that Hofstede's dimensions are among the most intensively investigated psychological constructs in the last three decades.

As labeled by Hofstede, the four dimensions pertain to power distance, individualism-collectivism, masculinity-femininity, and uncertainty avoidance, respectively. Subsequent to the completion of Hofstede's original data collection, a fifth dimension was added on the basis of a somewhat different methodology. It refers to short- versus long-term orientation and was originally designed to span the continuum between Confucian and Western dynamism. Power distance was defined by Hofstede and Hofstede as the extent to which unequal distribution of power is accepted as a normal or natural state of affairs. Individualism-collectivism is a bipolar dimension, bounded by the exaltation of the self at one extreme and the overriding prominence of the group or collectivity at the other. Masculinity-femininity refers to the characteristic overlap between gender roles that is encountered within a culture. Such overlap is less extensive in masculine and more extensive in feminine cultures. Moreover, masculine cultures are tilted toward achievement and performance, and feminine cultures favor caring. Uncertainty avoidance is high in those cultures where situations that lack clarity or structure are experienced as unpleasant or noxious. Low uncertainty-avoidance settings thrive on ambiguity, unpredictability, and improvisation. Long-term orientation differs from short-term orientation in the degree to which a culture's members are willing to postpone immediate reinforcement in favor of delayed rewards.

In the present context, it is worth noting that, as reported by Hofstede and Hofstede, the U.S. culture has been found to be the highest in individualism among the 72 cultures for which such indicators are available and that Russia's rank on this dimension is 37th and Japan's, 46th.[36] On the other hand, Russia clearly exceeds both Japan and the United States in power distance, in which it ranks 6th while Japanese and American ranks are 50th and 58th, respectively. In uncertainty avoidance, the ranks are 7th for Russia; for Japan, 12th; and for the United States, 63rd. In masculinity-femininity, Japan is more masculine, ranking 2nd, while the United

---

35. G. Hofstede, *Culture's Consequences: Comparing Values, Behaviors, Institutions and Organizations Across Nations*, 2nd ed. (Thousand Oaks, CA: Sage, 2001); G. Hofstede and G. J. Hofstede, *Cultures and Organizations: Software of the Mind*, 2nd ed. (New York: McGraw-Hill, 2005).

36. G. Hofstede and G. J. Hofstede, *Cultures and Organizations: Software of the Mind*.

States and Russia placed 19th and 63rd, respectively. Japan is also substantially higher in long-term orientation, placing 4th on this dimension, with United States coming in 29th out of 38 countries. Somewhat surprisingly, Russian university student samples were higher in long-term orientation than those of their counterparts in the United States, a finding that indicates persistent paternalism in the Russian culture. These findings, however, need to be extended and replicated.

To what extent (if any) and in what specific situations are these cultural value dimensions practically relevant to international spaceflight? We might expect that the demanding characteristics of spaceflight and the rigorous training that all astronauts undergo would override any cultural differences in Hofstede's dimensions. Moreover, we admit freely that the available cultural ranks for these traits may not be predictive of a specific individual's value pattern or of his or her performance. However, observations over several decades of international spaceflights forcefully show that cultural differences do matter. Extrapolation from the rapidly growing body of findings pertaining to personality differences between Russians and Americans, recently reviewed by Ritsher, strongly suggests that Hofstede's dimensions tie together many results.[37] In particular, the importance of collectivism-individualism has been widely recognized. In fact, Yuri Gagarin anticipated Hofstede's findings and formulations by stating in 1968 that "in our country, it is much easier to form a crew for a long-duration space mission than in capitalist countries [because we] are collectivist by nature."[38]

A. Merritt's comparison of airline pilots from 19 nations on Hofstede's dimensions found a significant effect of these variables on attitudes and behaviors, even though there was a detectable commonality across countries based on pilots' occupational culture. Differences across cultures were more pronounced when the indicators were made more relevant to the pilots' occupational context. Merritt concluded that these results "may extend to any population that is hierarchical in nature and involves teams of individuals interacting in high stress, high technology environments."[39]

---

37. Ritsher, "Cultural Factors in the International Space Station."
38. As quoted in Ritsher, "Cultural Factors": B135.
39. A. Merritt, "Culture in the Cockpit: Do Hofstede's Dimensions Replicate?" *Journal of Cross-Cultural Psychology* 31 (2000): 299.

OTHER APPROACHES TO CULTURALLY DISTINCTIVE VALUES

On a theoretical basis, S. H. Schwartz and A. Bardi have asserted that values constitute the central, nodal features of cultures that inevitably radiate to and affect all aspects of human behavior, conduct, and performance.[40] Schwartz has developed and extensively investigated in 20 nations a set of 11 values based on important and stable life goals.[41] Schwartz's value categories have been studied in relation to national differences (American versus Russian), majority versus minority status, and host versus guest status in multinational space crews. These results are reported in chapter 7 of this volume. On Earth, Russian teachers endorsed hierarchy and conservatism to a greater degree and intellectual and affective autonomy to a lesser degree than did their colleagues in several countries of Western Europe.

Relevant as the differences in collectivism-individualism are, they should not be the sole or principal focus of investigation. As already noted, even greater contrasts have been uncovered in power distance and uncertainty avoidance, as have substantial differences in masculinity-femininity and long-term orientation. Recent historical and sociological analyses of Russian society and polity emphasize the role of vertical and hierarchical, dominant-submissive relationships within the Russian social structure.[42] Their reverberations in human interaction, especially in the context of teamwork under demanding and stressful conditions, do not appear to have been studied systematically or intensively. Similarly, it is not yet clear in what ways, if at all, the substantial Russian-American differences in uncertainty avoidance may be manifested in space. The discrepancies in ratings in masculinity-femininity and long-term versus short-term orientation appear to be baffling or even counterintuitive and call for open-ended exploration through basic and applied research. In

---

40. S. H. Schwartz and A. Bardi, "Influences of Adaptation to Communist Rule on Value Priorities in Eastern Europe," *Political Psychology* 18 (1997): 385–410.

41. Shalom H. Schwartz, "Universals in the Content and Structure of Values: Theoretical Advances and Empirical Tests in 20 Countries," *Advances in Experimental Social Psychology* 25 (1992): 1–65.

42. Y. Afanas'ev, *Opasnaya Rossia (Dangerous Russia)* (Moscow: Russian State Humanities University Press, 2001); V. Shlapentokh, *Fear in Contemporary Society: Its Negative and Positive Effects* (New York: Palgrave Macmillan, 2006); V. Shlapentokh, *Contemporary Russia as a Feudal Society* (New York: Palgrave Macmillan, 2007).

what sense are the Russian society and its members committed to long-term goals, and what form does the feminine orientation take in social interaction, attitudes, and behavior? The initial impression is that of a conservative, authoritarian, though more matriarchal than patriarchal, society, which appears paradoxical in light of Russia's radical and revolutionary history through much of the 20th century.

Beyond these widely studied, internationally comparable dimensions, future investigators should open new lines of inquiry, proceeding from general observations or important themes in Russian culture. An example would be the systematic exploration of the role of patience, especially under adversity and other forms of stress, often described as an important mode of adaptation in Russia.[43]

EXPANDING THE RANGE OF CULTURES

As space exploration becomes ever more international, the approaches and findings of cross-cultural psychology may be extended to other cultures. Japanese astronauts have flown with both Americans and Russians. Observations and experiences from these flights, from both the hosts' and the Japanese participants' points of view, would provide valuable guidance in planning future spaceflights. There is substantial and diverse research-based literature on cultural characteristics in personality, emotional expression, and social behavior in Japan. Thus, in addition to Hofstede's dimensions in Japan, described earlier, one can draw upon studies of differences between Japanese and Americans and ways to bridge them; characteristics of Japanese selves, modes of interpersonal relating, and socialization within the family and at school; and many other themes.[44] Collectively, these contributions point to the prominence of group orientation in Japan that is reflected in self-experience, interpersonal relations, and styles of adaptation. But it would be an oversimplification to describe the Japanese self as sociocentric or the Japanese

---

43. V. Shlapentokh, "Russian Patience: A Reasonable Behavior and a Social Strategy," *Archives Euripeenes de Sociologie* 36 (1995): 247–280.

44. W. Goodykunst and T. Nishida, *Bridging Japanese/North American Differences* (Thousand Oaks, CA: Sage Publications, 1994); N. Rosenberger, ed., *Japanese Sense of Self* (Cambridge: Cambridge University Press, 1992); L. T. Doi, *Anatomy of Dependence* (Tokyo: Kodansha, 1973); D. W. Shwalb and B. J. Shwalb, eds., *Japanese Childrearing: Two Generations of Scholarship* (New York: Guilford Press, 1996).

social orientation as collectivistic. Instead, more than five decades of systematic research reveals a more complex picture pointing to an intricate intertwining of collectivistic attitudes with individualistic striving. Japanese and Americans both aspire toward positive selves. However, Heine, Takata, and Lehman have concluded that "Japanese self seems better characterized by a need to secure a positive view from others rather than from oneself and this securing of others' approval seems to be better served by self-improvement rather than self-enhancement."[45] Performance and well-being of the Japanese tend to be enhanced in groups composed of their compatriots; subjective and output effects of being a minority of one among strangers in space await systematic investigation.

CAUTIONS AND LIMITATIONS

In past decades, cultural factors in behavior and experience tended to be slighted or overlooked; with the current emphasis on cultural awareness, there is an emergent danger of exaggerating the importance of culture. The former tendency leads to cultural insensitivity; the latter trend promotes stereotyping and thereby reduces personal sensitivity. Let us forever keep in mind that differences between individuals who come from different cultures are not automatically traceable to culture. In complex situations encountered in the course of international spaceflight, culture is likely to interact with other factors in affecting behavior and well-being, and in some instances, culture may have had a minor or negligible influence. Moreover, astronauts and cosmonauts are of necessity exceptional individuals, certainly shaped by their respective cultures, but also statistically discrepant from most of their cultural peers. We face a formidable undertaking to disentangle the cultural threads operating in any concrete instance. It is a challenge that requires persistence, patience, and creativity to formulate general principles of culture's influence in interaction with the person and his or her context. Increasing participation of female astronauts from various nations will provide new challenges and opportunities for studying and optimizing the culture by gender interactions in space.

---

45. S. J. Heine, T. Takata, and D. R. Lehman, "Beyond Self-Presentation: Evidence for Self-Criticism Among Japanese," *Personality and Social Psychology Bulletin* 26 (2000): 71.

INTERCULTURAL COOPERATION IN SPACE: A UNIQUE SOURCE OF CROSS-CULTURAL DATA

The preceding sections have dealt with the potential contributions of cross-cultural psychology to the implementation of international missions. We now turn the tables and see what unique intercultural interaction in space may contribute to the store of knowledge of cross-cultural psychology. Observations from space should be thoroughly and factually documented. Given the singularity of the situations involved, qualitative observations of interactions and encounters should lead the way. Whenever possible, they should be quantified and studied by means of procedures for N of 1, followed by the gradual imposition of formal but flexible research designs.[46]

Qualitative retrospective study of the evolution of international missions should include interviews of the surviving veterans of the early stages of space exploration, with emphasis on the social and personal values and philosophies of life as well as ideologies and beliefs that animated these concerns. How have these national and personal goals been transformed in the course of the last 50 years, as the cultures of both the United States and Russia have undergone various degrees of change, from the social upheaval in America in the 1960s and 1970s to the demise of the Soviet political and economic system?

With only slight exaggeration, adaptation to space missions, especially long-duration missions, can be likened to adaptation to a small society or a mini-culture, with its own internal rules and evolving customs as well as externally imposed imperatives. What, then, are the characteristics of the communities of space exploration on the ground and in space, and what are their commonalities across nations and their culturally specific hallmarks?

Proceeding from these considerations, we should identify the characteristics of spaceflight participants and the cultural differences, if any, between astronauts of participating nations. Equally importantly, researchers might find that astronauts' similarities may override national characteristics. Perhaps a shared core of traits accounts for the effective and productive, though by no means problem-free, outcome of joint missions in space.

---

46. W. F. Dukes, "N=1," *Psychological Bulletin* 64 (1965): 74–79.

Greater availability of participant observations of spacefarers from numerous nations is one of the byproducts of the modern space exploration era. Spacefarers' perspectives on the dynamics of monocultural and bicultural space crews provide a major resource that should be systematically tapped and utilized. Conversely, the accounts of the subjective experiences of these guest astronauts are a source of information on being a minority of one in a novel, demanding, and stressful undertaking.

Space exploration has been both a challenge and a boon to cross-cultural psychology: a challenge by demonstrating that, even though it is a young branch of its discipline, it has methods and findings to contribute to the continued progress of space exploration, and a boon by making it possible to collect a wealth of unique data of cultural relevance. These data remain to be sifted, categorized, and incorporated into a set of general principles. Potentially, cross-cultural psychology stands to be tremendously enriched as this happens. We hope that cross-cultural psychology's contribution to international missions will constitute a partial repayment.

Afterword
From the Past to the Future

Gro Mjeldheim Sandal
   Department of Psychosocial Science
   University of Bergen

Gloria R. Leon
   Department of Psychology
   University of Minnesota

ABSTRACT

Although NASA has been criticized for many years for neglecting psychological issues in research and operations, the past several years have witnessed an increased recognition of the importance of psychosocial and cultural factors in the success and safety of human space missions. The challenges associated with future long-duration missions involving extreme environments, isolation, and greater crew autonomy as the distance from Mission Control increases require effective countermeasures to mitigate the risk for behavioral health problems, psychiatric disorders, and impairments in effective crew interactions and task performance. International space missions also underscore the need to understand the potential safety implications of individual and cultural differences at a national, organizational, and professional level that involve both space crews and ground-based personnel. While the research literature on space psychology has increased over the last few years, many unanswered questions remain that require additional investigation.

INTRODUCTION

Since the first solo flight of a human being into Earth orbit, human spaceflight has undergone significant changes in terms of crew composition, mission duration, and complexity. Even the major achievement of the establishment of the

International Space Station (ISS), put into service in 2000, must be regarded as just one further step toward a much bigger endeavor: human exploratory missions into outer space and the establishment of human outposts on other celestial bodies of the solar system. This effort might include a return to the Moon and the establishment of a lunar station for permanent occupation, as well as flights of humans to our neighbor planet Mars. Technology is just one important aspect of such long-duration space missions. Beyond that, there is no doubt that different biomedical and psychological factors might represent important limitations to the success of the missions.

The chapters presented in this volume and publications elsewhere demonstrate advances in our understanding of challenges related to human spaceflights. Yet a number of issues remain that require additional investigation. This is particularly the case in regard to long-duration exploratory space missions to the Moon and Mars, which to some extent can be expected to involve the same range of psychological issues and risks that have been reported from long-duration orbital flights, simulation studies, and expeditions into analog environments. Nonetheless, missions to Mars will add a new dimension to the history of human expeditions into *terrae incognitae* with respect to the distance and duration of travel. Such missions might not be comparable to any other undertaking humans have ever attempted because of the long distance of travel, the duration of constant dependence upon automated life-support systems, the degree of isolation and confinement, and the lack of short-term rescue possibilities in case of emergencies. Current knowledge about human adaptation under these conditions is very limited, but predictions about the emergence of certain psychological issues can be extrapolated from Earth-based analogs and studies, as well as previous spaceflights. For example, historical expeditions to unknown parts of the world parallel some of the human requirements associated with future interplanetary missions. In addition, studies of personnel wintering over on Antarctic research stations, such as the Concordia station, in which the European Space Agency (ESA) has been involved in recent years, may provide highly useful data in such fields as telemedicine as well as psychophysiological reactions and group dynamics associated with isolation and lack of evacuation possibilities. A 500-day space simulation study in hyperbaric chambers is currently under way to model the living conditions of crews on future Mars missions. While these earthbound analogs provide interesting platforms for research, a much more detailed understanding of the concrete scientific and operational demands of planetary exploration mission space crews and the design of their habitats is needed before the psychological issues associated with these missions can finally be assessed.

It is beyond doubt that political and commercial interests, to a large extent, will dictate crew composition for future missions, and these forces might introduce new factors that must be addressed as part of mission planning. An era of space tourism seems to be at hand, as highlighted by the study presented by Harvey Wichman in chapter 5. Now that the Russian, European, and American space agencies are equal partners on joint projects such as the ISS, they are faced with challenges related to collaboration among people with different cultural backgrounds. For a long time, the impact of cultural variability seemed to be a neglected area in aerospace research. As demonstrated by several chapters in this volume, the last few years have witnessed an increased recognition of the potential safety risks associated with cultural heterogeneity in terms of nationality, organizational background, and profession. Accidents like those involving *Challenger* and *Columbia* have underscored the need for future studies to focus not only on people operating in space, but also on ground-based personnel.

Management of the safety culture may become even more complex in international space programs in which mission preparation and control often demand the coordinated effort of different space agencies. An interesting study by L. Tomi, P. Stefanowska, and V. F. Doyle involved data collection among ISS personnel from different agencies.[1] The results demonstrated the need for training and intervention beyond the space crews and the need to address differences in organizational cultures, in addition to those between national cultures. A similar conclusion was reached in another study involving 600 employees working for ESA that addressed challenges associated with collaboration with representatives from other agencies. Preliminary results indicate that the most prominent difficulties differed according to agency and seem to reflect value differences anchored in organizational and national culture.[2]

---

1   L. Tomi, P. Stefanowska, and V. F. Doyle, "Cross-cultural Training Requirements Definition Survey for the International Space Station" (paper presented at the 115th Annual Convention of the American Psychological Association, San Francisco, CA, August 2007).

2   G. M. Sandal and D. Manzey, "Cultural Determinants of Co-working of Ground Personnel in the European Space Agency," *Acta Astronautica* 65 (2009): 1520–1569.

In contrast to the interest traditionally shown by the Russians in space psychology, an interest now shared by European as well as Japanese space agencies, NASA has been criticized for neglecting psychological issues in research and operations. Many researchers and others have observed that the NASA culture discouraged questions about the behavioral health of astronauts since they were assumed to have "the right stuff." Recently, these issues have gained more attention. In chapter 2, Albert Harrison and Edna Fiedler describe some recent positive developments within the NASA establishment in moving away from "the marginalization of psychology." A more realistic recognition of stress and its consequences has led to a concern with prevention and countermeasures as a result of experiences during the Russian-American Shuttle-*Mir* Program and the ISS.

Greater attention to the psychological health of astronauts is reflected in a NASA-sponsored behavioral health workshop held several years ago, followed by some increase in more formal operational attention to postmission functioning of the astronaut and his/her family. However, there is still much planning, program implementation, and change in organizational culture that needs to be carried out for these efforts to be more than a perfunctory gesture. A major problem in terms of astronaut and perhaps, to some extent, family behavioral health is the organizational factors that work against disclosure of psychological problems. The concern by astronauts that disclosure will result in nonassignment or removal from a flight assignment likely has considerable justification in reality. The "right stuff" culture and concerns about confidentiality feed into a system where personal problems might be denied or dealt with in dysfunctional ways such as substance abuse, impulsive or high-risk behaviors, and family and other interpersonal conflicts. The findings of the 2007 NASA Astronaut Health Care System Review Committee indicated that none of those interviewed knew of an astronaut who had sought behavioral health care.

A focus on behavioral assessment and health should extend from the process of selection to training, in-flight, and postflight periods. Passing a psychological/psychiatric screening upon entrance to the astronaut corps does not predict the absence of psychological/behavioral problems that might occur at a later time. In considering crew selection for future long-duration lunar and planetary missions, a particularly thorny issue is the prediction of the later development of psychopathology in initially healthy individuals. This situation would not only affect the disabled person, but would also have a detrimental effect on overall crew performance and safety—potentially, it could even jeopardize the mission. Some disorders, for example, certain types of depression and anxiety disorders, as well as liability for schizophrenia, have a considerable genetic component that might be detected by inquiries about family background. As the field of genetic testing and other statistical predictors of dysfunction develop over time, this information could be used in screening, selection, and later evaluation. However, in fairness to individual astronauts, it is very important to remain cognizant of the fact that "liability" is not the same as the development of a disorder. Moreover, in reality, it seems impossible to deal with possible deleterious behavioral health consequences in space simply by focusing on selection criteria, which thus far have been the most prominent countermeasure applied by space agencies. Therefore, the careful development and implementation of psychological, pharmaceutical, and other countermeasures to deal with these potential problems is extremely important.

In recent years, a number of reports, papers, and books have identified and emphasized psychological problems related to long-duration spaceflights. Concerns have been raised about the stressors to which crewmembers are exposed and the potential deleterious effects on health, group functioning, performance, and safety. More recently, Peter Suedfeld has brought a new perspective into the field by emphasizing the salutogenic, growth-enhancing aspects of experiences in space as well as in other extreme environments and conditions. Rather than focusing on pathogenesis, this approach directs attention to the fostering of human strengths that act as buffers against psychological dysfunction and the adverse effects of stress. Chapter 7, by Suedfeld, Kasia Wilk, and Lindi Cassel, presents a careful study of the change in values and coping mechanisms by majority and minority crewmembers from preflight to in-flight and postflight status, with somewhat different trajectories and value changes between the two groups. The in-flight increase by minority members in spirituality and family interest reflects positive growth; the decline in references to power and self-direction may be a reflection of a "host/guest" phenomenon. While these value changes were not sustained postflight, it is possible that with the experience of an extended flight to the Moon or Mars, greater changes in values would occur that would be more stable and would extend to the postflight period. In general, a transformation of a value hierarchy might represent a coping strategy for maintaining motivation during long-duration missions. This transformation might also be reflected in terms of certain personality characteristics becoming more prominent or diminished, depending on the individual's space experiences.

The salutogenic aspects of space are also reflected in the study of self-initiated photography during leisure time in space and the larger number of photos taken as the mission progressed, described in chapter 4 by Julie Robinson and colleagues. Many astronauts have commented on the awesome, existential phenomenon of viewing Earth from space, and it is the primary leisure-time activity. The perhaps "hard-wired" pleasure of communing with nature and the feeling of spirituality that is often a part of this activity seem to form an important positive component of life in space; they are also a means of coping with negative aspects of space missions. For example, Russian cosmonauts have commented on the pleasure they experienced on space missions by growing plants, watering and tending to them and watching the plants thrive. A challenge for future long-duration missions to distant planets where Earth and home will no longer be in view is to plan for other methods of viewing the cosmos and opportunities for leisure photography during planetary exploration.

While the transformation of values may be regarded as a constructive coping mechanism, these changes might also carry potential hazardous aspects. During a future expedition to Mars or beyond, a partial or complete loss of commitment to the usual (earthbound) system of values and behavioral norms could result that, in extreme cases, might involve unforeseeable risks in terms of individual behavior, performance of mission tasks, and interpersonal interactions within the crew. This might make any external control and guidance of the crew extremely difficult. Experiences from some military combat units indicate that microcultures that develop in isolated groups can diverge profoundly in values and behaviors from societal norms (e.g., the groupthink phenomenon). An open-minded investigation might be necessary to understand group processes that are likely to occur in confined, isolated, and autonomous crews.

Any breakdown of communication, cooperation, and cohesiveness of a space crew must be considered as an important limiting factor to mission success and safety. One concern raised by several of the authors in this volume is that impairments of crew interactions and operations might be induced by cultural and gender differences. With long-duration missions, one would anticipate changes and fluctuations in interpersonal as well as task cohesion over time and the possible influence of factors such as gender and culture on these relationships. Chapter 6, by Jason Kring and Megan Kaminski, addresses the issue of gender composition and crew cohesion, also examining unidimensional and multidimensional conceptions and types of cohesion, for example, interpersonal versus task cohesion. While there is a growing literature on mixed-gender and multicultural groups in polar settings and space simulations, the information from space missions is primarily anecdotal. With respect to gender, this is partly because of the low number of women involved and the associated need to maintain confidentiality. Also, the small sample sizes make it difficult to isolate the effects of gender or culture from factors such as personality and professional training experiences. Most spaceflight experiences with multicultural space crews, thus far, stem from international Shuttle or *Mir* missions where cross-cultural aspects were inevitably confounded with "host/guest" differences, i.e., where the missions usually involved one "dominant culture" being the host for crewmembers from other countries, organizations, and professions.

There is no doubt that national interests will continue to play a major role in decisions about the cultural mix of crews for specific missions. Within that context, decisions about crew composition ideally should be made on the basis of the compatibility of a group of persons possessing the range of skills required for the particular mission. This information can be obtained by behavioral observations of a larger group from which a particular crew will be selected. The use of a uniform assessment battery across countries to provide data to inform about selection for specific space missions would also be helpful. However, in reality, this information might be difficult to obtain. Space agencies have shown differences in preferences regarding measurements. Additionally, the same measure might not be equally applicable or valid for different national groups. In the context of long-duration spaceflight such as a future Mars mission, there is still very little empirical evidence to inform on the ideal crew mix despite the many discussions that have ensued. Addressing this issue by conducting systematic research in analog environments will require a significant commitment by space agencies in terms of priorities and funding. However, it is likely that existing basic research and studies from other fields on broader populations can provide important insights into group processes in spaceflights. For example, information about cooperation in multinational groups is available from fields such as anthropology, sociology, and cross-cultural psychology. Although one needs to be mindful about the fact that astronauts represent a highly selected group, findings from broader populations can, to a large extent, be used to generate empirical testable hypotheses for research in analog environments.

A comprehensive program in spaceflight behavioral health must be broad based; be interdisciplinary; and address issues at the individual, small group, and organizational levels. In dealing with issues of relevance to multinational spaceflight, cross-cultural collaboration in research should be considered mandatory to prevent ethnocentrism in design and interpretations. Such collaboration, although often challenging, might provide research results that are more easily accepted and applicable across national agencies. Rigorous theories and thorough methodological designs are prerequisites for progress in the field. In chapter 3, Sheryl Bishop points out that one key methodological and validity issue is the added value of utilizing consistent measures across various expeditions, allowing a more valid comparison of individuals and teams across environments, including space. With awareness of this problem, researchers in the last several years have tended more often to use a common set of measures that facilitate the comparison of results across studies. While this is a positive development, potential cultural bias in the assessment methodology must not be overlooked. For example, J. B. Ritsher has emphasized that cultural differences may affect the accuracy of methods for detecting distress in spaceflights, so specific methods will not work equivalently.[3]

Robust and sensitive assessment methods for monitoring behavior and health in space are crucial for obtaining a high quality of research, but also for the early detection of behavioral health problems. Improved prediction, prevention, and treatment of distress will improve the safety of international long-duration space missions. With regard to both prevention and treatment, the development of countermeasures designed for autonomous crews become more important as we prepare for much longer exploration-class missions to Mars and beyond. One example is the development of a computer-interactive video countermeasure technology for the prevention and treatment of depression, and another is a program for conflict resolution under the leadership of Dr. James Carter, Harvard–Beth Israel Deaconess Medical Center. This countermeasure has a number of features that appear to make it quite acceptable for astronaut use on a mission—using the astronaut's familiarity and comfort with computer technologies, supplying confidentiality because the astronaut can work through the program in the privacy of his or her quarters, teaching coping methods, and focusing on prevention and early intervention to avoid having problems spiral out of control. Evaluation and comparisons of this and other innovative countermeasures in multinational contexts represent impor-

---

3. J. B. Ritsher, "Cultural Factors and the International Space Station," *Aviation, Space, and Environmental Medicine* 76, no. 6, supplement (2005): 135–144.

tant next steps that need to be taken in the future. In our view, giving a high priority to research programs in the area of behavioral health will prove to be highly beneficial for the safety of long-duration space missions.

About the Authors

Pamela J. Baskin graduated with her B.A. in biological sciences from the University of Houston, Clear Lake, in 1997. She is currently working on her master's degree in industrial/organizational psychology at the University of Houston, Clear Lake. She serves as the Lead Research Coordinator in the Behavioral Health and Performance (BHP) research area at JSC and also serves as the Human Research Program Analog Project Coordinator. She has assisted in the coordination of BHP's involvement with three NASA Extreme Environment Mission Operations (NEEMO) missions. She is currently assisting with the coordination of BHP's involvement in the 2008 Haughton-Mars Project and the 105-Day Russian Chamber Study. She can be reached at *pamela.baskins-1@nasa.gov*, telephone 281-212-1360, or at Wyle Integrated Science and Engineering Group, 1290 Hercules Drive, Houston, TX 77058.

Sheryl L. Bishop graduated with her B.A. in psychology from the University of Texas at San Antonio in 1982 and her Ph.D. in social psychology from the University of Texas at Austin in 1989. She joined the faculty of the School of Nursing at the University of Texas Medical Branch (UTMB) at Galveston in 1992 as a biostatistician. In 1995, she joined the School of Medicine at UTMB as a member of the departments of Health and Safety Services, Family Medicine, and eventually Preventive Medicine and Community Health, where she served as the academic coordinator for graduate programs. She advanced to associate professor in 2000 and has been instrumental in the development of the NASA/UTMB Space Medicine Fellowship, the Aerospace Medicine Residency program, and the Space Life Sciences Ph.D. curriculum. Bishop returned to the School of Nursing as graduate faculty and senior biostatistician in 2007. She is also a full faculty member at the International Space University, Strasbourg, France, contributing to the Summer Session Program since 1994 and the Master's of Space Science Program since its inception. Since 1990, Bishop has investigated human performance and group dynamics in teams in extreme environments, including deep cavers, mountain climbers, desert survival groups, polar expeditioners, and Antarctic winter-over groups, and in numerous simulations of isolated, confined environments for space. She has participated in various television documentaries and has over 60 publications and over 50 scholarly presentations in both the medical and psychological fields on topics as diverse as psychometric assessment, research methodology, outcomes research, psychosocial group dynamics, and human performance in extreme environments. She is a senior associate editor and editorial board member for the Society for Human Performance in Extreme Environments and its journal, *Human Performance in Extreme Environments*, as well as the *International Journal of Earth Science, Life Support and Biosphere Science*. She serves as review editor for *Aviation, Space, and Environmental Medicine* and *Astra Astronautica* and as grants reviewer for the Canadian Space Agency. She can be reached at *sbishop@utmb.edu*, telephone 409-772-8216, or at the School of Nursing, Rte. 1029, University of Texas Medical Branch, Galveston, TX 77555.

Jennifer E. Boyd graduated with distinction from Stanford University with an A.B. in psychology in 1989. She completed her M.A. and Ph.D. in clinical/community psychology at the University of Maryland, graduating in 1997. She finished a three-year postdoctoral fellowship in the Psychiatric Epidemiology Training program at Columbia University in 2000. She worked as a researcher at the Center for Health Care Evaluation at the Palo Alto Veterans Affairs (VA) and Stanford University Medical Center from 2000 to 2002. In 2002, she joined the faculty at the University of California, San Francisco. She is the founding director of the Psychosocial Rehabilitation and Recovery Center at the San Francisco VA Medical Center. Her research focuses on the influence of sociocultural factors on psychopathology and its detection, such as the influence of stigma on the course of psychosis, the effect of isolated and confined environments on mental health, and the effect of culture on symptom patterns and the accuracy of assessment instruments. She can be reached at *jennifer.boyd@ucsf.edu*, telephone 415-221-4810, or at the Department of Psychiatry, UCSF/VA Medical Center, 4150 Clement Street (116A), San Francisco, CA 94121.

Lindi Cassel graduated from the University of British Columbia in 2005 with a B.S. in psychology. Following graduation, she worked as a research assistant under Dr. Peter Suedfeld, conducting research on diverse topics such as resilience in genocide survivors, terrorism, political psychology, and personality factors contributing to successful human spaceflight. Her research has been published in the *Journal of Positive Psychology* and *Psychology Today*. Cassel is completing her M.Sc. in occupational therapy at McMaster University and can be contacted at *cassellj@mcmaster.ca*.

Juris G. Draguns is professor emeritus of psychology at Pennsylvania State University. Born in Latvia, he started his primary schooling in his native country, graduated from high school in Germany, and completed his university studies in the United States. He holds a B.A. degree from Utica College and a Ph.D. in clinical psychology from the University of Rochester. His first two professional appointments were as clinical and research psychologist at Rochester, New York, and Worcester, Massachusetts, State Hospitals. He also lectured at the University of Rochester and Clark University. For 30 years, he was a faculty member of Pennsylvania State University as an associate and a full professor of psychology. Retired as professor emeritus in 1997, he continues to be professionally active through private practice, writing, lecturing, and research.

Draguns is author or coauthor of over 160 monographs, articles, and chapters in psychology, psychiatry, and anthropology. He coedited *Counseling Across Cultures* (in six editions); *Defense Mechanisms: Theoretical, Research, and Clinical Perspectives* (in two editions); *Handbook of Culture, Therapy, and Healing*; *Personality and Person Perception Across Cultures*; *The Roots of Perception*; and *Psychological Processes in Cognition and Personality*.

Throughout his career, Draguns has cultivated international and cross-cultural contacts and has developed a network of international collaborators. He has taught and lectured, in six languages, in Germany, Sweden, Australia, Switzerland, Taiwan, Latvia, and Mexico and has presented papers at congresses and conferences in 29 countries.

After the demise of the Soviet Union, Draguns participated in developing undergraduate and graduate programs in psychology in Latvia. In 2006, he took part in a site visit to the Psychology Department of the University of Kuwait. In 2001, he received the American Psychological Association's Award for Distinguished Contributions to the International Advancement of Psychology. In 2003, the University of Latvia bestowed upon him the degree of doctor honoris causa; in 2005, he was elected president of the Society for Cross-Cultural Research; and in

2008, he received the Emeritus Distinction Award from the College of Liberal Arts of Pennsylvania State University.

Draguns can be reached at the Department of Psychology, Pennsylvania State University, University Park, PA 16802, or at *jgd1@psu.edu*, telephone 814-238-1504, or fax 814-863-7002.

Edna R. Fiedler received her B.S. degree from Trinity University, San Antonio, Texas, and her M.A. and Ph.D. in social psychology from the University of Colorado. Subsequently, she went on for further training in clinical child psychology. She joined the faculty of the Department of Psychology, St. Mary's University, San Antonio, Texas, rising to the rank of professor and department chair. In 1989, she became director of an Air Force–wide screening program that screened all incoming recruits for psychopathology incompatible with a career in the military. She also was research director for the Neuropsychiatric Division of Wilford Hall, Lackland Air Force Base. From 1998 to 2003, Fiedler was a branch manager at the Civil Aerospace Medical Institute, Federal Aviation Agency (FAA), Oklahoma City, Oklahoma. The FAA position involved working closely with various national and international counterparts and committees. Currently, she serves as the Health and Science Liaison for the National Space Biomedical Research Institute (NSBRI) and is a faculty member of the Psychiatry Department, Baylor College of Medicine. In this role, she coordinates efforts among NSBRI team leads and NASA's research and operational elements. Fiedler regularly publishes in peer-reviewed journals and has served on numerous academic and research committees. Her honors include teaching and job performance awards at St. Mary's University, the FAA, and NASA. She is a member of the American Psychological Association and the American Psychological Society. She can be reached at *efiedler@bcm.edu* or NSBRI, Suite NA 425, Baylor College of Medicine, Houston, TX 77030.

Albert A. Harrison received his B.A. and M.A. in psychology from the University of California, Santa Barbara, and his Ph.D. in social psychology from the University of Michigan. In 1967, he joined the faculty of the Department of Psychology at the University of California, Davis, and in 1979, he advanced to professor. Now professor emeritus, he is the author or coauthor of approximately

100 papers in a wide range of journals, and his books include *Living Aloft: Human Requirements for Extended Spaceflight*, *From Antarctica to Outer Space: Life in Isolation and Confinement*, *After Contact: The Human Response to Extraterrestrial Life*, and *Spacefaring: The Human Dimension*. His most recent book, *Starstruck: Cosmic Visions in Science, Religion, and Folklore*, describes how rapidly accumulating scientific findings about our place in the universe are encouraging people to seek new answers to old existential questions.

Harrison is a member of the Permanent SETI (Search for Extraterrestrial Intelligence) Study Group of the International Academy of Astronautics, and he was a member of NASA's Space Human Factors Engineering Science and Technology Working Group. In December 2003, he was Principal Investigator of a NASA-sponsored conference on new directions in behavioral health, and he edited a special supplement on this topic for *Aviation, Space, and Environmental Medicine*. He is currently involved in planetary defense (protecting Earth from asteroids and comets) and is heavily involved in the International Academy of Astronautics' Space Architecture Study Group, seeking new approaches to human-centered design.

A former deputy U.S. editor of *Systems Research and Behavioral Science*, he may be reached at *aaharrison@ucdavis.edu*, telephone 530-756-2361, or at the Department of Psychology, One Shields Avenue, University of California Davis, Davis, CA 95616.

Megan A. Kaminski received her B.S. in human factors psychology from Embry-Riddle Aeronautical University and is currently in the master's program in Human Factors and Applied Cognition at George Mason University. She has presented at the 2007 and 2008 Florida Student Conference for Human Factors and Applied Psychology. She can be reached at *meganannakaminski@yahoo.com*.

Jason P. Kring received his B.A. in psychology from the University of Colorado, Boulder; his M.S. in experimental psychology from Emporia State University; and his Ph.D. in applied experimental and human factors psychology from the University of Central Florida. He has worked as a researcher at the United States Army Research Institute in Orlando and interned at NASA's

Johnson Space Center in Houston, where he developed human factors recommendations related to the International Space Station. He is currently an assistant professor of human factors and systems at Embry-Riddle Aeronautical University in Daytona Beach, Florida.

Kring serves as president of the Society for Human Performance in Extreme Environments, an interdisciplinary forum for scientists, operational personnel, and students with an interest and expertise in the area of human performance and behavior in complex, high-stress environments. He also serves as editor of the *Journal of Human Performance in Extreme Environments* and is codirector of the Team Simulation and Gaming Laboratory at Embry-Riddle.

Kring has over 30 publications and 50 conference presentations representing a wide range of research interests including performance in extreme and stressful settings; the effects of cohesion, trust, and communication on team performance; crew composition and interactions for long-duration spaceflight; aerospace human factors; distributed team performance; and training applications of simulation and virtual reality.

Gloria R. Leon, Ph.D., is professor emeritus of psychology in the Department of Psychology at the University of Minnesota as of June 2006. Prior to that, she participated in both the clinical and personality doctoral training areas of the department and was an adjunct professor in the Institute of Child Development at the University of Minnesota. She received her doctorate in mental health psychology from the University of Maryland. She was appointed assistant professor in the Department of Psychology at the University of Minnesota in 1974; in addition to her teaching and research activities, she served as assistant, associate, and, for 10 years, director of the clinical psychology graduate program. She has conducted extensive space analog research on personality, behavioral functioning, and team processes of different polar expedition groups, studying teams composed of single-gender, mixed-gender, and cross-national members. Leon has also carried out research on groups who have experienced traumatic situations, including Vietnam combat and nurse veterans and those impacted by the Chernobyl disaster. Leon continues to be active in research and national and international activities in space and disaster areas. She has been a member of several NASA committees and workshops focused on behavioral health and human performance in space, NASA peer review panels, the International Academy of Astronautics' psychosocial committee, and program committees for various space-related congresses. From 2004 to 2007, she was a member of the external advisory council of the NSBRI, advising the Neurobehavioral and Psychosocial team, and currently serves on several Institute of Medicine committees dealing with the health of astronauts and others in extreme environments. She is a board member and chair of the World Association for Disaster Medicine's international psychosocial task force to develop guidelines for responding to and managing disasters, and she is on the editorial board of *Prehospital and Disaster Medicine*.

Valerie A. Olson is a doctoral candidate in social/cultural anthropology at Rice University. She is currently completing an ethnographic dissertation on contemporary American astronautics culture, after conducting fieldwork at NASA and the ESA and within the private space development sector. Using coordinated case studies on space biomedicine, space analogs, space architecture, planetary science, and interplanetary mission design, her dissertation specifically analyzes how astronautics scientists and engineers understand and produce knowledge about the human-environment relationship. From 2005 through 2007, she served as an NSBRI research intern in support of behavioral health research. She can be reached at *valeriao@rice.edu*.

Julie A. Robinson is the Program Scientist for the ISS at NASA JSC. She serves as the chief scientist for the ISS Program, representing all ISS research inside and outside the Agency. She provides recommendations regarding research on the ISS to the ISS Program Manager and the Space Operations and Exploration Systems Mission Directorates at NASA Headquarters.

Robinson has an interdisciplinary background in the physical and biological sciences. Her professional experience has included research activities in a variety of fields, including virology, analytical chemistry, genetics, statistics, field biology, and remote sensing. She has authored over 50 scientific publications. She earned a B.S. in chemistry and a B.S. in biology from Utah State University in 1989. She earned a Ph.D. in ecology, evolution, and conservation biology from the University of Nevada, Reno, in 1996.

She began her career in the Earth Science and Image Analysis Laboratory at JSC (working for Lockheed Martin), serving as the Earth Science Applications Lead. In this work, she collaborated with numerous ecologists and conservation biologists in incorporating remote sensing data into their projects. She led the development of the *Gateway to Astronaut Photography of Earth* on the World Wide Web at *http://eol.jsc.nasa.gov*, which distributes hundreds of thousands of images of Earth taken from orbit to scientists and the public each month. In 2004, she completed a major NASA-sponsored scientific project to develop global maps of coral reefs from Landsat 7 data and facilitate a distribution network to make them accessible to natural resource managers. She was an editor of the book *Dynamic Earth Environments: Remote Sensing Observations from Shuttle*-*Mir Missions* and continues work on a remote sensing textbook.

In 2004, she joined NASA as a civil servant in the Office of the ISS Program Scientist, where she led the development of a system for tracking ISS research and results and provided the information to the public via the NASA Web portal. She was named Deputy ISS Program Scientist in 2006 and Program Scientist in 2007.

Gro Mjeldheim Sandal is a professor of psychology at the University of Bergen in Norway. Since the early 1990s, she has been the Principal Investigator of large-scale international research projects funded by the ESA and focused on psychological reactions during human spaceflights. These projects have included a number of simulation studies of multinational crews isolated in hyperbaric chambers and personnel operating in other extreme environments (Antarctic research stations, polar crossings, and military settings). She is currently leading one of the first psychological studies of resident crews on the ISS in collaboration with colleagues working for the Russian Space Agency. Her recent research has focused on the implications of individual and cultural differences in values for efficient coworking among crews in space as well as among ground-based personnel. A major aim is to gain knowledge that can be used for selection, training, and in-flight support.

Kelley J. Slack is an industrial/organizational psychologist working for Wyle in the BHP group for NASA at JSC. As part of BHP since 2000, she is involved in the psychological and psychiatric selection process for astronauts and

About the Authors

psychological preparedness and support of ISS astronauts. Her current research involves Antarctica as an analog for long-duration spaceflight and competencies necessary for successful Moon and Mars missions.

Slack attended the London School of Economics before receiving her B.A. from Rice University. She received her M.A. in psychology and her Ph.D. in industrial/organizational psychology with a minor in statistics from the University of Houston.

She may be reached at *kslack@wylehou.com*, telephone 281-212-1404, or at Wyle Integrated Science and Engineering Group, 1290 Hercules Drive, Houston, TX 77058.

Peter Suedfeld was born in Hungary and immigrated to the United States after World War II. After serving in the U.S. Army, he received his B.A. from Queens College of the City University of New York and his M.A. and Ph.D. in experimental psychology from Princeton University. He taught at the University of Illinois and Rutgers University prior to joining the University of British Columbia in 1972 as head of the Department of Psychology. He later became dean of the Faculty of Graduate Studies, and he now holds emeritus status.

His research, published in seven books and over 270 book chapters and journal articles, has focused on the strengths of people as they cope during and after experiencing extreme, unusual, challenging, and traumatic events and environments. His methodology has included laboratory experiments in profound stimulus reduction; fieldwork in the Antarctic and the High Arctic; and interview and questionnaire studies with Holocaust survivors, prisoners in solitary confinement, and astronauts. More recently, he has been developing and applying methods of quantitative content analysis to archival material produced by individuals in those groups as well as solitary sailors, early explorers, mountain climbers, and high-level political and military leaders in situations of personal, national, and international stress.

Suedfeld is a Fellow of the Royal Society of Canada (the National Academy), the Canadian Psychological Association (president, 1990–91), the American Psychological Association (six divisions), the Academy of Behavioral Medicine Research, and other scientific organizations. He is a corresponding member of the International Academy of Astronautics, and he is the only psychologist elected as an Honorary Fellow of the Royal Canadian Geographical Society. He has received the highest awards for scientific contributions from the Canadian Psychological Association and the International Society of Political Psychology, as well as the Antarctica Service Medal of the National Science Foundation and the Zachor Award of the Parliament of Canada for contributions to Canadian society. He has served on many advisory and consultative committees in the United States and Canada and has chaired the Canadian Antarctic Research Program and the Life Sciences Advisory Committee of the Canadian Space Agency.

Suedfeld can be reached at *psuedfeld@psych.ubc.ca*, telephone 604-822-5713, or at the Department of Psychology, The University of British Columbia, 2136 West Mall, Vancouver, BC V6T 1Z4, Canada.

Michael H. Trenchard graduated with his B.S. in meteorology from Texas A&M University in 1973 and his M.S. in meteorology/climatology from Texas A&M. He has worked as a government contractor in the fields of meteorology and remote sensing in various NASA remote sensing programs at both Johnston Space Center and Stennis Space Center, with an emphasis on agricultural and commercial applications. Since 1995, he has worked in Crew Earth Observations supporting both the Shuttle and International Space Station Programs. Activities have included weather satellite imagery interpretation for on-orbit operations in crew photography; crew training in weather- and climate-related observations from space; and, more recently, the cataloging and interpretation of high-resolution, hand-held imagery acquired by crews with digital camera systems. He is a member of the American Meteorological Society and may be reached at *mike.trenchard-1@nasa.gov*, telephone 281-483-5333.

Douglas A. Vakoch received his B.A. in religion from Carleton College, his M.A. in history and philosophy of science from the University of Notre Dame, and his M.A. and Ph.D. in clinical psychology from the State University of New York at Stony Brook. After finishing his predoctoral clinical internship in health psychology at the University of Wisconsin-Madison Hospital and Clinics, he completed a National Research Service Award postdoctoral fellowship in the Department of Psychology at Vanderbilt University. Following this fellowship, Vakoch joined the

About the Authors

SETI Institute, a nonprofit research organization in Silicon Valley, where he is now the director of Interstellar Message Composition. He is also a professor in the Department of Clinical Psychology at the California Institute of Integral Studies. He is the editor of several forthcoming volumes, including *Between Worlds: The Art and Science of Interstellar Message Composition*; *Ecofeminism and Rhetoric: Critical Perspectives on Sex, Technology, and Discourse*; and (with Albert A. Harrison) *Civilizations beyond Earth: Extraterrestrial Life and Society*.

Vakoch is chair of the International Academy of Astronautics' Study Groups on Interstellar Message Construction and Active SETI, and as a member of the International Institute of Space Law, he examines policy issues related to interstellar communication. He may be reached at *dvakoch@seti.org*, telephone 650-960-4514, or at the Center for SETI Research, SETI Institute, 189 Bernardo Avenue, Suite 100, Mountain View, CA 94043; or at *dvakoch@ciis.edu*, telephone 415-575-6244, or at the Department of Clinical Psychology, California Institute of Integral Studies, 1453 Mission Street, San Francisco, CA 94103.

Harvey Wichman received his B.A. and M.A. degrees from California State University, Long Beach, and his Ph.D. in experimental psychology from Claremont Graduate University. He was a member of the founding faculties of both Delta College in Michigan and California State University in San Bernardino. He is professor emeritus at Claremont McKenna College (CMC) and Claremont Graduate University.

Trained in both neuroscience and social psychology, he is a graduate of NASA's Biospace Technology Training Program at Wallops Island, Virginia. He conducts research on the effects of working and living in extreme environments. As a Fellow of the American Council on Education, he spent a year at the National Institutes of Health. As a Sloan Foundation Fellow, he spent a year designing the International Space Station with the team at Rockwell International in Downey, California, following which he served as a Faculty Fellow at the Jet Propulsion Laboratory in Pasadena, California. He is the author of the book *Human Factors in the Design of Spacecraft*, and he has published articles in journals such as the *Journal of Personality and Social Psychology*, *Space Life Sciences*, *Human Factors*, and *Aviation, Space, and Environmental Medicine*. As director of CMC's Aerospace Psychology Laboratory, he has conducted space research involving designing passenger compartments for civilian spaceflight on reusable McDonnell Douglas Aerospace (now Boeing Aerospace) rockets for both orbital and suborbital flights. The CMC Aerospace Psychology Laboratory has conducted research in simulated spaceflights and recently designed a spaceflight simulator for the space museum at Alamogordo, New Mexico. He may be reached at *harvey@wichman.org*, telephone 909-607-7986, or at Claremont Graduate University, Harper Hall East, B5, 160 East Tenth Street, Claremont, CA 91711.

Kasia E. Wilk graduated from the University of British Columbia with her B.Sc. in behavioral neuroscience, and she completed her M.Sc. degree in consciousness and transpersonal psychology at John Moores University in Liverpool, England. She has studied psychological effects amongst astronaut and cosmonaut crewmembers during outer space missions in the Reactions to Environmental Stress and Trauma (REST) laboratory of Dr. Peter Suedfeld in the Department of Psychology at the University of British Columbia; she also has worked in other areas of psychology research including twin personality traits and the effects of conditioned epileptical seizures in rats. She is planning on pursuing a Ph.D. in transpersonal psychology and hopes to work with various humanitarian organizations that help children in difficult environments. She is currently a researcher at the Youth Forensic Psychiatric Services Research and Evaluation Department under the Ministry of Children and Family Development in Burnaby, British Columbia, studying the effectiveness of provincial treatment programs used in rehabilitating young offenders.

Kimberly J. Willis graduated with her B.S. in physical sciences and a concentration in geology from the University of Houston, Clear Lake, in 1982, and she was granted her M.S. in 1986. She also earned an M.S. in secondary education from the University of Houston, Clear Lake, in 1990 and holds a Lifetime Texas Provisional Certification in secondary education, Earth science, and physical science. She joined the Earth Observations group in 1990 and was responsible for the metadata assembly of the Earth-viewing photography taken by Space Shuttle astronauts. In 1991, she became the task lead for metadata assembly of astronaut photography of Earth. By 1996, Willis had conducted program planning, operations, and data analysis for the NASA/*Mir* Earth Observations Program. She also established and documented operational procedures for the Crew Earth Observations (CEO) on the ISS. At the same time, she participated in Space Shuttle mission planning, astronaut training in

About the Authors

Earth science disciplines, and real-time Shuttle operations. She advanced to Project Manager, Earth Observations, in 2001. In 2004, Willis developed and coordinated all Earth science training plans and materials for Space Shuttle and Space Station astronauts. Her duties included the management of all aspects of Earth science training and Earth science operations for the Space Shuttle and the ISS. Today, Willis is the Section Manager over Astromaterials Research and Curation, which includes the CEO group. Along with her administrative tasks, she is the task lead for CEO. Willis is also adjunct faculty at the University of Houston, Clear Lake, where she teaches fundamentals of Earth science. She can be reached at *kim.willis-1@nasa.gov*, telephone 281-244-1215.

Acronyms and Abbreviations

AIAA	American Institute of Aeronautics and Astronautics
APA	American Psychological Association
BCPR	Bioastronautics Critical Path Roadmap
BHP	Behavioral Health and Performance
CEO	Crew Earth Observations
CIIS	California Institute of Integral Studies
CMC	Claremont McKenna College
CSA	Canadian Space Agency
EPICA	European Project for Ice Coring in Antarctica
ESA	European Space Agency
ESCG	Engineering and Science Contract Group (NASA JSC)
ESTEC	European Space Research and Technology Centre
EU	European Union
EUE	extreme unusual environment
EVA	extravehicular activity
FAA	Federal Aviation Agency
FMARS	Flashline Mars Arctic Research Station
FY	fiscal year
GENMOD	procedure for generalized linear models in SAS
GLIMMIX	generalized linear mixed models
HMP	Haughton-Mars Project
HRP	Human Research Program
IAA	International Academy of Astronautics
IGY	International Geophysical Year
IKI	Russian Institute for Space Research
IRP	Integrated Research Plan
ISS	International Space Station
JSC	Johnson Space Center
kph	kilometers per hour
LDM	Long Duration Mission
LDSF	long-duration spaceflight
LEAPS	Long-term Effects After Prolonged Spaceflight
LIWC	Linguistic Inquiry and Word Count

MDRS	Mars Desert Research Station
NASA	National Aeronautics and Space Administration
NEEMO	NASA Extreme Environment Mission Operations
NEO-PI	Neuroticism, Extraversion, and Openness to Experience–Personality Inventory
NOAA	National Oceanic and Atmospheric Administration
ns	not significant
NSBRI	National Space Biomedical Research Institute
NURC	National Undersea Research Center
PDA	personal digital assistant
REST	Reactions to Environmental Stress and Trauma
SAE	Society of Automotive Engineers
SBRI	Space Biomedical Research Institute
SETI	Search for Extraterrestrial Intelligence
SFINCSS	Simulation of Flight of International Crew on Space Station
SIOP	Society for Industrial and Organizational Psychologists
SSEOP	Space Shuttle Earth Observations Project
SUNY	State University of New York
TCA	Thematic Content Analysis
U.K.	United Kingdom
UCLA	University of California, Los Angeles
UTMB	University of Texas Medical Branch
VA	Veterans Affairs
WET-F	Weightless Environmental Training Facility
WISE	Women in Space Earliest

The NASA History Series

REFERENCE WORKS
NASA SP-4000

Grimwood, James M. *Project Mercury: A Chronology*. NASA SP-4001, 1963.

Grimwood, James M., and Barton C. Hacker, with Peter J. Vorzimmer. *Project Gemini Technology and Operations: A Chronology*. NASA SP-4002, 1969.

Link, Mae Mills. *Space Medicine in Project Mercury*. NASA SP-4003, 1965.

Astronautics and Aeronautics, 1963: Chronology of Science, Technology, and Policy. NASA SP-4004, 1964.

Astronautics and Aeronautics, 1964: Chronology of Science, Technology, and Policy. NASA SP-4005, 1965.

Astronautics and Aeronautics, 1965: Chronology of Science, Technology, and Policy. NASA SP-4006, 1966.

Astronautics and Aeronautics, 1966: Chronology of Science, Technology, and Policy. NASA SP-4007, 1967.

Astronautics and Aeronautics, 1967: Chronology of Science, Technology, and Policy. NASA SP-4008, 1968.

Ertel, Ivan D., and Mary Louise Morse. *The Apollo Spacecraft: A Chronology, Volume I, Through November 7, 1962*. NASA SP-4009, 1969.

Morse, Mary Louise, and Jean Kernahan Bays. *The Apollo Spacecraft: A Chronology, Volume II, November 8, 1962–September 30, 1964*. NASA SP-4009, 1973.

Brooks, Courtney G., and Ivan D. Ertel. *The Apollo Spacecraft: A Chronology, Volume III, October 1, 1964–January 20, 1966*. NASA SP-4009, 1973.

Ertel, Ivan D., and Roland W. Newkirk, with Courtney G. Brooks. *The Apollo Spacecraft: A Chronology, Volume IV, January 21, 1966–July 13, 1974*. NASA SP-4009, 1978.

Astronautics and Aeronautics, 1968: Chronology of Science, Technology, and Policy. NASA SP-4010, 1969.

Newkirk, Roland W., and Ivan D. Ertel, with Courtney G. Brooks. *Skylab: A Chronology*. NASA SP-4011, 1977.

Van Nimmen, Jane, and Leonard C. Bruno, with Robert L. Rosholt. *NASA Historical Data Book, Volume I: NASA Resources, 1958–1968.* NASA SP-4012, 1976, rep. ed. 1988.

Ezell, Linda Neuman. *NASA Historical Data Book, Volume II: Programs and Projects, 1958–1968.* NASA SP-4012, 1988.

Ezell, Linda Neuman. *NASA Historical Data Book, Volume III: Programs and Projects, 1969–1978.* NASA SP-4012, 1988.

Gawdiak, Ihor, with Helen Fedor. *NASA Historical Data Book, Volume IV: NASA Resources, 1969–1978.* NASA SP-4012, 1994.

Rumerman, Judy A. *NASA Historical Data Book, Volume V: NASA Launch Systems, Space Transportation, Human Spaceflight, and Space Science, 1979–1988.* NASA SP-4012, 1999.

Rumerman, Judy A. *NASA Historical Data Book, Volume VI: NASA Space Applications, Aeronautics and Space Research and Technology, Tracking and Data Acquisition/Support Operations, Commercial Programs, and Resources, 1979–1988.* NASA SP-4012, 1999.

Rumerman, Judy A. *NASA Historical Data Book, Volume VII: NASA Launch Systems, Space Transportation, Human Spaceflight, and Space Science, 1989–1998.* NASA SP-2009-4012, 2009.

No SP-4013.

Astronautics and Aeronautics, 1969: Chronology of Science, Technology, and Policy. NASA SP-4014, 1970.

Astronautics and Aeronautics, 1970: Chronology of Science, Technology, and Policy. NASA SP-4015, 1972.

Astronautics and Aeronautics, 1971: Chronology of Science, Technology, and Policy. NASA SP-4016, 1972.

Astronautics and Aeronautics, 1972: Chronology of Science, Technology, and Policy. NASA SP-4017, 1974.

Astronautics and Aeronautics, 1973: Chronology of Science, Technology, and Policy. NASA SP-4018, 1975.

Astronautics and Aeronautics, 1974: Chronology of Science, Technology, and Policy. NASA SP-4019, 1977.

Astronautics and Aeronautics, 1975: Chronology of Science, Technology, and Policy. NASA SP-4020, 1979.

Astronautics and Aeronautics, 1976: Chronology of Science, Technology, and Policy. NASA SP-4021, 1984.

Astronautics and Aeronautics, 1977: Chronology of Science, Technology, and Policy. NASA SP-4022, 1986.

Astronautics and Aeronautics, 1978: Chronology of Science, Technology, and Policy. NASA SP-4023, 1986.

Astronautics and Aeronautics, 1979–1984: Chronology of Science, Technology, and Policy. NASA SP-4024, 1988.

Astronautics and Aeronautics, 1985: Chronology of Science, Technology, and Policy. NASA SP-4025, 1990.

Noordung, Hermann. *The Problem of Space Travel: The Rocket Motor.* Edited by Ernst Stuhlinger and J. D. Hunley, with Jennifer Garland. NASA SP-4026, 1995.

Astronautics and Aeronautics, 1986–1990: A Chronology. NASA SP-4027, 1997.

Astronautics and Aeronautics, 1991–1995: A Chronology. NASA SP-2000-4028, 2000.

Orloff, Richard W. *Apollo by the Numbers: A Statistical Reference.* NASA SP-2000-4029, 2000.

Lewis, Marieke, and Ryan Swanson. *Astronautics and Aeronautics: A Chronology, 1996–2000.* NASA SP-2009-4030, 2009.

Ivey, William Noel, and Ryan Swanson. *Astronautics and Aeronautics: A Chronology, 2001–2005.* NASA SP-2010-4031, 2010.

MANAGEMENT HISTORIES
NASA SP-4100

Rosholt, Robert L. *An Administrative History of NASA, 1958–1963.* NASA SP-4101, 1966.

Levine, Arnold S. *Managing NASA in the Apollo Era.* NASA SP-4102, 1982.

Roland, Alex. *Model Research: The National Advisory Committee for Aeronautics, 1915–1958.* NASA SP-4103, 1985.

Fries, Sylvia D. *NASA Engineers and the Age of Apollo.* NASA SP-4104, 1992.

Glennan, T. Keith. *The Birth of NASA: The Diary of T. Keith Glennan.* Edited by J. D. Hunley. NASA SP-4105, 1993.

Seamans, Robert C. *Aiming at Targets: The Autobiography of Robert C. Seamans.* NASA SP-4106, 1996.

Garber, Stephen J., ed. *Looking Backward, Looking Forward: Forty Years of Human Spaceflight Symposium.* NASA SP-2002-4107, 2002.

Mallick, Donald L., with Peter W. Merlin. *The Smell of Kerosene: A Test Pilot's Odyssey.* NASA SP-4108, 2003.

Iliff, Kenneth W., and Curtis L. Peebles. *From Runway to Orbit: Reflections of a NASA Engineer.* NASA SP-2004-4109, 2004.

Chertok, Boris. *Rockets and People, Volume I.* NASA SP-2005-4110, 2005.

Chertok, Boris. *Rockets and People: Creating a Rocket Industry, Volume II.* NASA SP-2006-4110, 2006.

Chertok, Boris. *Rockets and People: Hot Days of the Cold War, Volume III.* NASA SP-2009-4110, 2009.

Laufer, Alexander, Todd Post, and Edward Hoffman. *Shared Voyage: Learning and Unlearning from Remarkable Projects.* NASA SP-2005-4111, 2005.

Dawson, Virginia P., and Mark D. Bowles. *Realizing the Dream of Flight: Biographical Essays in Honor of the Centennial of Flight, 1903–2003.* NASA SP-2005-4112, 2005.

Mudgway, Douglas J. *William H. Pickering: America's Deep Space Pioneer.* NASA SP-2008-4113, 2008.

PROJECT HISTORIES
NASA SP-4200

Swenson, Loyd S., Jr., James M. Grimwood, and Charles C. Alexander. *This New Ocean: A History of Project Mercury.* NASA SP-4201, 1966; rep. ed. 1999.

Green, Constance McLaughlin, and Milton Lomask. *Vanguard: A History.* NASA SP-4202, 1970; rep. ed. Smithsonian Institution Press, 1971.

Hacker, Barton C., and James M. Grimwood. *On the Shoulders of Titans: A History of Project Gemini.* NASA SP-4203, 1977; rep. ed. 2002.

Benson, Charles D., and William Barnaby Faherty. *Moonport: A History of Apollo Launch Facilities and Operations.* NASA SP-4204, 1978.

Brooks, Courtney G., James M. Grimwood, and Loyd S. Swenson, Jr. *Chariots for Apollo: A History of Manned Lunar Spacecraft.* NASA SP-4205, 1979.

The NASA History Series

Bilstein, Roger E. *Stages to Saturn: A Technological History of the Apollo/Saturn Launch Vehicles.* NASA SP-4206, 1980 and 1996.

No SP-4207.

Compton, W. David, and Charles D. Benson. *Living and Working in Space: A History of Skylab.* NASA SP-4208, 1983.

Ezell, Edward Clinton, and Linda Neuman Ezell. *The Partnership: A History of the Apollo-Soyuz Test Project.* NASA SP-4209, 1978.

Hall, R. Cargill. *Lunar Impact: A History of Project Ranger.* NASA SP-4210, 1977.

Newell, Homer E. *Beyond the Atmosphere: Early Years of Space Science.* NASA SP-4211, 1980.

Ezell, Edward Clinton, and Linda Neuman Ezell. *On Mars: Exploration of the Red Planet, 1958–1978.* NASA SP-4212, 1984.

Pitts, John A. *The Human Factor: Biomedicine in the Manned Space Program to 1980.* NASA SP-4213, 1985.

Compton, W. David. *Where No Man Has Gone Before: A History of Apollo Lunar Exploration Missions.* NASA SP-4214, 1989.

Naugle, John E. *First Among Equals: The Selection of NASA Space Science Experiments.* NASA SP-4215, 1991.

Wallace, Lane E. *Airborne Trailblazer: Two Decades with NASA Langley's 737 Flying Laboratory.* NASA SP-4216, 1994.

Butrica, Andrew J., ed. *Beyond the Ionosphere: Fifty Years of Satellite Communications.* NASA SP-4217, 1997.

Butrica, Andrew J. *To See the Unseen: A History of Planetary Radar Astronomy.* NASA SP-4218, 1996.

Mack, Pamela E., ed. *From Engineering Science to Big Science: The NACA and NASA Collier Trophy Research Project Winners.* NASA SP-4219, 1998.

Reed, R. Dale. *Wingless Flight: The Lifting Body Story.* NASA SP-4220, 1998.

Heppenheimer, T. A. *The Space Shuttle Decision: NASA's Search for a Reusable Space Vehicle.* NASA SP-4221, 1999.

Hunley, J. D., ed. *Toward Mach 2: The Douglas D-558 Program.* NASA SP-4222, 1999.

Swanson, Glen E., ed. *"Before This Decade Is Out . . ." Personal Reflections on the Apollo Program.* NASA SP-4223, 1999.

Tomayko, James E. *Computers Take Flight: A History of NASA's Pioneering Digital Fly-By-Wire Project.* NASA SP-4224, 2000.

Morgan, Clay. *Shuttle-Mir: The United States and Russia Share History's Highest Stage.* NASA SP-2001-4225, 2001.

Leary, William M. *"We Freeze to Please": A History of NASA's Icing Research Tunnel and the Quest for Safety.* NASA SP-2002-4226, 2002.

Mudgway, Douglas J. *Uplink-Downlink: A History of the Deep Space Network, 1957–1997.* NASA SP-2001-4227, 2001.

No SP-4228 or SP-4229.

Dawson, Virginia P., and Mark D. Bowles. *Taming Liquid Hydrogen: The Centaur Upper Stage Rocket, 1958–2002.* NASA SP-2004-4230, 2004.

Meltzer, Michael. *Mission to Jupiter: A History of the Galileo Project.* NASA SP-2007-4231, 2007.

Heppenheimer, T. A. *Facing the Heat Barrier: A History of Hypersonics.* NASA SP-2007-4232, 2007.

Tsiao, Sunny. *"Read You Loud and Clear!" The Story of NASA's Spaceflight Tracking and Data Network.* NASA SP-2007-4233, 2007.

CENTER HISTORIES
NASA SP-4300

Rosenthal, Alfred. *Venture into Space: Early Years of Goddard Space Flight Center.* NASA SP-4301, 1985.

Hartman, Edwin P. *Adventures in Research: A History of Ames Research Center, 1940–1965.* NASA SP-4302, 1970.

Hallion, Richard P. *On the Frontier: Flight Research at Dryden, 1946–1981.* NASA SP-4303, 1984.

Muenger, Elizabeth A. *Searching the Horizon: A History of Ames Research Center, 1940–1976.* NASA SP-4304, 1985.

Hansen, James R. *Engineer in Charge: A History of the Langley Aeronautical Laboratory, 1917–1958.* NASA SP-4305, 1987.

Dawson, Virginia P. *Engines and Innovation: Lewis Laboratory and American Propulsion Technology.* NASA SP-4306, 1991.

Dethloff, Henry C. *"Suddenly Tomorrow Came . . .": A History of the Johnson Space Center, 1957–1990.* NASA SP-4307, 1993.

Hansen, James R. *Spaceflight Revolution: NASA Langley Research Center from Sputnik to Apollo*. NASA SP-4308, 1995.

Wallace, Lane E. *Flights of Discovery: An Illustrated History of the Dryden Flight Research Center*. NASA SP-4309, 1996.

Herring, Mack R. *Way Station to Space: A History of the John C. Stennis Space Center*. NASA SP-4310, 1997.

Wallace, Harold D., Jr. *Wallops Station and the Creation of an American Space Program*. NASA SP-4311, 1997.

Wallace, Lane E. *Dreams, Hopes, Realities. NASA's Goddard Space Flight Center: The First Forty Years*. NASA SP-4312, 1999.

Dunar, Andrew J., and Stephen P. Waring. *Power to Explore: A History of Marshall Space Flight Center, 1960–1990*. NASA SP-4313, 1999.

Bugos, Glenn E. *Atmosphere of Freedom: Sixty Years at the NASA Ames Research Center*. NASA SP-2000-4314, 2000.

No SP-4315.

Schultz, James. *Crafting Flight: Aircraft Pioneers and the Contributions of the Men and Women of NASA Langley Research Center*. NASA SP-2003-4316, 2003.

Bowles, Mark D. *Science in Flux: NASA's Nuclear Program at Plum Brook Station, 1955–2005*. NASA SP-2006-4317, 2006.

Wallace, Lane E. *Flights of Discovery: An Illustrated History of the Dryden Flight Research Center*. NASA SP-2007-4318, 2007. Revised version of NASA SP-4309.

Arrighi, Robert S. *Revolutionary Atmosphere: The Story of the Altitude Wind Tunnel and the Space Power Chambers*. NASA SP-2010-4319, 2010.

GENERAL HISTORIES
NASA SP-4400

Corliss, William R. *NASA Sounding Rockets, 1958–1968: A Historical Summary*. NASA SP-4401, 1971.

Wells, Helen T., Susan H. Whiteley, and Carrie Karegeannes. *Origins of NASA Names*. NASA SP-4402, 1976.

Anderson, Frank W., Jr. *Orders of Magnitude: A History of NACA and NASA, 1915–1980*. NASA SP-4403, 1981.

Sloop, John L. *Liquid Hydrogen as a Propulsion Fuel, 1945–1959.* NASA SP-4404, 1978.

Roland, Alex. *A Spacefaring People: Perspectives on Early Spaceflight.* NASA SP-4405, 1985.

Bilstein, Roger E. *Orders of Magnitude: A History of the NACA and NASA, 1915–1990.* NASA SP-4406, 1989.

Logsdon, John M., ed., with Linda J. Lear, Jannelle Warren Findley, Ray A. Williamson, and Dwayne A. Day. *Exploring the Unknown: Selected Documents in the History of the U.S. Civil Space Program, Volume I: Organizing for Exploration.* NASA SP-4407, 1995.

Logsdon, John M., ed., with Dwayne A. Day and Roger D. Launius. *Exploring the Unknown: Selected Documents in the History of the U.S. Civil Space Program, Volume II: External Relationships.* NASA SP-4407, 1996.

Logsdon, John M., ed., with Roger D. Launius, David H. Onkst, and Stephen J. Garber. *Exploring the Unknown: Selected Documents in the History of the U.S. Civil Space Program, Volume III: Using Space.* NASA SP-4407, 1998.

Logsdon, John M., ed., with Ray A. Williamson, Roger D. Launius, Russell J. Acker, Stephen J. Garber, and Jonathan L. Friedman. *Exploring the Unknown: Selected Documents in the History of the U.S. Civil Space Program, Volume IV: Accessing Space.* NASA SP-4407, 1999.

Logsdon, John M., ed., with Amy Paige Snyder, Roger D. Launius, Stephen J. Garber, and Regan Anne Newport. *Exploring the Unknown: Selected Documents in the History of the U.S. Civil Space Program, Volume V: Exploring the Cosmos.* NASA SP-2001-4407, 2001.

Logsdon, John M., ed., with Stephen J. Garber, Roger D. Launius, and Ray A. Williamson. *Exploring the Unknown: Selected Documents in the History of the U.S. Civil Space Program, Volume VI: Space and Earth Science.* NASA SP-2004-4407, 2004.

Logsdon, John M., ed., with Roger D. Launius. *Exploring the Unknown: Selected Documents in the History of the U.S. Civil Space Program, Volume VII: Human Spaceflight: Projects Mercury, Gemini, and Apollo.* NASA SP-2008-4407, 2008.

Siddiqi, Asif A., *Challenge to Apollo: The Soviet Union and the Space Race, 1945–1974.* NASA SP-2000-4408, 2000.

Hansen, James R., ed. *The Wind and Beyond: Journey into the History of Aerodynamics in America, Volume 1: The Ascent of the Airplane.* NASA SP-2003-4409, 2003.

Hansen, James R., ed. *The Wind and Beyond: Journey into the History of Aerodynamics in America, Volume 2: Reinventing the Airplane.* NASA SP-2007-4409, 2007.

Hogan, Thor. *Mars Wars: The Rise and Fall of the Space Exploration Initiative.* NASA SP-2007-4410, 2007.

MONOGRAPHS IN AEROSPACE HISTORY
NASA SP-4500

Launius, Roger D., and Aaron K. Gillette, comps. *Toward a History of the Space Shuttle: An Annotated Bibliography.* Monographs in Aerospace History, No. 1, 1992.

Launius, Roger D., and J. D. Hunley, comps. *An Annotated Bibliography of the Apollo Program.* Monographs in Aerospace History, No. 2, 1994.

Launius, Roger D. *Apollo: A Retrospective Analysis.* Monographs in Aerospace History, No. 3, 1994.

Hansen, James R. *Enchanted Rendezvous: John C. Houbolt and the Genesis of the Lunar-Orbit Rendezvous Concept.* Monographs in Aerospace History, No. 4, 1995.

Gorn, Michael H. *Hugh L. Dryden's Career in Aviation and Space.* Monographs in Aerospace History, No. 5, 1996.

Powers, Sheryll Goecke. *Women in Flight Research at NASA Dryden Flight Research Center from 1946 to 1995.* Monographs in Aerospace History, No. 6, 1997.

Portree, David S. F., and Robert C. Trevino. *Walking to Olympus: An EVA Chronology.* Monographs in Aerospace History, No. 7, 1997.

Logsdon, John M., moderator. *Legislative Origins of the National Aeronautics and Space Act of 1958: Proceedings of an Oral History Workshop.* Monographs in Aerospace History, No. 8, 1998.

Rumerman, Judy A., comp. *U.S. Human Spaceflight: A Record of Achievement, 1961–1998.* Monographs in Aerospace History, No. 9, 1998.

Portree, David S. F. *NASA's Origins and the Dawn of the Space Age.* Monographs in Aerospace History, No. 10, 1998.

Logsdon, John M. *Together in Orbit: The Origins of International Cooperation in the Space Station.* Monographs in Aerospace History, No. 11, 1998.

Phillips, W. Hewitt. *Journey in Aeronautical Research: A Career at NASA Langley Research Center*. Monographs in Aerospace History, No. 12, 1998.

Braslow, Albert L. *A History of Suction-Type Laminar-Flow Control with Emphasis on Flight Research*. Monographs in Aerospace History, No. 13, 1999.

Logsdon, John M., moderator. *Managing the Moon Program: Lessons Learned from Apollo*. Monographs in Aerospace History, No. 14, 1999.

Perminov, V. G. *The Difficult Road to Mars: A Brief History of Mars Exploration in the Soviet Union*. Monographs in Aerospace History, No. 15, 1999.

Tucker, Tom. *Touchdown: The Development of Propulsion Controlled Aircraft at NASA Dryden*. Monographs in Aerospace History, No. 16, 1999.

Maisel, Martin, Demo J. Giulanetti, and Daniel C. Dugan. *The History of the XV-15 Tilt Rotor Research Aircraft: From Concept to Flight*. Monographs in Aerospace History, No. 17, 2000. NASA SP-2000-4517.

Jenkins, Dennis R. *Hypersonics Before the Shuttle: A Concise History of the X-15 Research Airplane*. Monographs in Aerospace History, No. 18, 2000. NASA SP-2000-4518.

Chambers, Joseph R. *Partners in Freedom: Contributions of the Langley Research Center to U.S. Military Aircraft of the 1990s*. Monographs in Aerospace History, No. 19, 2000. NASA SP-2000-4519.

Waltman, Gene L. *Black Magic and Gremlins: Analog Flight Simulations at NASA's Flight Research Center*. Monographs in Aerospace History, No. 20, 2000. NASA SP-2000-4520.

Portree, David S. F. *Humans to Mars: Fifty Years of Mission Planning, 1950–2000*. Monographs in Aerospace History, No. 21, 2001. NASA SP-2001-4521.

Thompson, Milton O., with J. D. Hunley. *Flight Research: Problems Encountered and What They Should Teach Us*. Monographs in Aerospace History, No. 22, 2001. NASA SP-2001-4522.

Tucker, Tom. *The Eclipse Project*. Monographs in Aerospace History, No. 23, 2001. NASA SP-2001-4523.

Siddiqi, Asif A. *Deep Space Chronicle: A Chronology of Deep Space and Planetary Probes, 1958–2000*. Monographs in Aerospace History, No. 24, 2002. NASA SP-2002-4524.

Merlin, Peter W. *Mach 3+: NASA/USAF YF-12 Flight Research, 1969–1979*. Monographs in Aerospace History, No. 25, 2001. NASA SP-2001-4525.

The NASA History Series

Anderson, Seth B. *Memoirs of an Aeronautical Engineer: Flight Tests at Ames Research Center: 1940–1970.* Monographs in Aerospace History, No. 26, 2002. NASA SP-2002-4526.

Renstrom, Arthur G. *Wilbur and Orville Wright: A Bibliography Commemorating the One-Hundredth Anniversary of the First Powered Flight on December 17, 1903.* Monographs in Aerospace History, No. 27, 2002. NASA SP-2002-4527.

No monograph 28.

Chambers, Joseph R. *Concept to Reality: Contributions of the NASA Langley Research Center to U.S. Civil Aircraft of the 1990s.* Monographs in Aerospace History, No. 29, 2003. NASA SP-2003-4529.

Peebles, Curtis, ed. *The Spoken Word: Recollections of Dryden History, The Early Years.* Monographs in Aerospace History, No. 30, 2003. NASA SP-2003-4530.

Jenkins, Dennis R., Tony Landis, and Jay Miller. *American X-Vehicles: An Inventory—X-1 to X-50.* Monographs in Aerospace History, No. 31, 2003. NASA SP-2003-4531.

Renstrom, Arthur G. *Wilbur and Orville Wright: A Chronology Commemorating the One-Hundredth Anniversary of the First Powered Flight on December 17, 1903.* Monographs in Aerospace History, No. 32, 2003. NASA SP-2003-4532.

Bowles, Mark D., and Robert S. Arrighi. *NASA's Nuclear Frontier: The Plum Brook Research Reactor.* Monographs in Aerospace History, No. 33, 2004. NASA SP-2004-4533.

Wallace, Lane, and Christian Gelzer. *Nose Up: High Angle-of-Attack and Thrust Vectoring Research at NASA Dryden, 1979–2001.* Monographs in Aerospace History, No. 34, 2009. NASA SP-2009-4534.

Matranga, Gene J., C. Wayne Ottinger, Calvin R. Jarvis, and D. Christian Gelzer. *Unconventional, Contrary, and Ugly: The Lunar Landing Research Vehicle.* Monographs in Aerospace History, No. 35, 2006. NASA SP-2004-4535.

McCurdy, Howard E. *Low-Cost Innovation in Spaceflight: The History of the Near Earth Asteroid Rendezvous (NEAR) Mission.* Monographs in Aerospace History, No. 36, 2005. NASA SP-2005-4536.

Seamans, Robert C., Jr. *Project Apollo: The Tough Decisions.* Monographs in Aerospace History, No. 37, 2005. NASA SP-2005-4537.

Lambright, W. Henry. *NASA and the Environment: The Case of Ozone Depletion.* Monographs in Aerospace History, No. 38, 2005. NASA SP-2005-4538.

Chambers, Joseph R. *Innovation in Flight: Research of the NASA Langley Research Center on Revolutionary Advanced Concepts for Aeronautics.* Monographs in Aerospace History, No. 39, 2005. NASA SP-2005-4539.

Phillips, W. Hewitt. *Journey into Space Research: Continuation of a Career at NASA Langley Research Center.* Monographs in Aerospace History, No. 40, 2005. NASA SP-2005-4540.

Rumerman, Judy A., Chris Gamble, and Gabriel Okolski, comps. *U.S. Human Spaceflight: A Record of Achievement, 1961–2006.* Monographs in Aerospace History, No. 41, 2007. NASA SP-2007-4541.

Dick, Steven J., Stephen J. Garber, and Jane H. Odom. *Research in NASA History.* Monographs in Aerospace History, No. 43, 2009. NASA SP-2009-4543.

Merlin, Peter W. *Ikhana: Unmanned Aircraft System Western States Fire Missions.* Monographs in Aerospace History, No. 44, 2009. NASA SP-2009-4544.

Fisher, Steven C., and Shamim A. Rahman. *Remembering the Giants: Apollo Rocket Propulsion Development.* Monographs in Aerospace History, No. 45, 2009. NASA SP-2009-4545.

ELECTRONIC MEDIA
NASA SP-4600

Remembering Apollo 11: The 30th Anniversary Data Archive CD-ROM. NASA SP-4601, 1999.

Remembering Apollo 11: The 35th Anniversary Data Archive CD-ROM. NASA SP-2004-4601, 2004. This is an update of the 1999 edition.

The Mission Transcript Collection: U.S. Human Spaceflight Missions from Mercury Redstone 3 to Apollo 17. NASA SP-2000-4602, 2001.

Shuttle-Mir: The United States and Russia Share History's Highest Stage. NASA SP-2001-4603, 2002.

U.S. Centennial of Flight Commission Presents Born of Dreams—Inspired by Freedom. NASA SP-2004-4604, 2004.

Of Ashes and Atoms: A Documentary on the NASA Plum Brook Reactor Facility. NASA SP-2005-4605, 2005.

Taming Liquid Hydrogen: The Centaur Upper Stage Rocket Interactive CD-ROM. NASA SP-2004-4606, 2004.

Fueling Space Exploration: The History of NASA's Rocket Engine Test Facility DVD. NASA SP-2005-4607, 2005.

Altitude Wind Tunnel at NASA Glenn Research Center: An Interactive History CD-ROM. NASA SP-2008-4608, 2008.

A Tunnel Through Time: The History of NASA's Altitude Wind Tunnel. NASA SP-2010-4609, 2010.

CONFERENCE PROCEEDINGS
NASA SP-4700

Dick, Steven J., and Keith Cowing, eds. *Risk and Exploration: Earth, Sea and the Stars.* NASA SP-2005-4701, 2005.

Dick, Steven J., and Roger D. Launius. *Critical Issues in the History of Spaceflight.* NASA SP-2006-4702, 2006.

Dick, Steven J., ed. *Remembering the Space Age: Proceedings of the 50th Anniversary Conference.* NASA SP-2008-4703, 2008.

Dick, Steven J., ed. *NASA's First 50 Years: Historical Perspectives.* NASA SP-2010-4704, 2010.

SOCIETAL IMPACT
NASA SP-4800

Dick, Steven J., and Roger D. Launius. *Societal Impact of Spaceflight.* NASA SP-2007-4801, 2007.

Dick, Steven J., and Mark L. Lupisella. *Cosmos and Culture: Cultural Evolution in a Cosmic Context.* NASA SP-2009-4802, 2009.

Subject Index

Able (monkey), 18–19
Activities
    docking, 80, 87, 90, 93, 146
    extravehicular activities (EVA), 4, 80, 87, 90, 93, 95, 150
    self-initiated, 80, 85, 87–88, 90–100
    training, 77, 81–83, 147, 149, 151, 153–154, 173
Adaptation factors, 8, 26, 36, 43, 53–54, 59, 61, 67, 71–73
    automation complexity, 56
    cohesion, 14, 56, 58, 67, 154
    confinement, 2, 49, 52, 54–55, 58–59, 61, 66–67, 69, 71–73, 75, 169
    confusion, 56, 72
    coping, 52, 57, 59–60, 70, 73, 144, 160–161, 166–167, 169, 171, 174, 199
    crew heterogeneity, 56
    cultural differences, 56, 63–64, 152, 170, 180–182, 188–191, 193–194, 197
    differential situational reality, 65–66
    fatigue, 56, 70, 72
    group, 3, 8, 11, 14, 17, 22–23, 152–153, 157, 159, 167, 169–170, 172–173
    individual, 198
    interpersonal conflict, 9, 50, 56, 58–59, 63–65
    isolation, 2, 49, 52, 54–55, 58–59, 61, 67, 70–73, 75, 153, 168–169
    language differences, 56
    microsociety, 9, 66
    motivation, 56, 60, 67, 70, 146, 156
    pathology, 59
    personality, 53, 56–57, 60–61, 151, 154–155, 174
    physical danger, 56
    privacy, 58
    returnability, 65, 198
    risk, 47, 52, 54, 56, 62, 66, 77
    social isolation, 54, 58, 168
    stress, 52–53, 56–57, 60, 70, 73, 147–148, 169, 171, 173, 198
    subjectivization of time, 65–66
    withdrawal, 27, 58
Aerospace medicine, 8
Aerospace Psychology Laboratory, 109, 122
Affirmative Action, 38–41. *See also* Equal Employment Opportunity
Agreeableness, 60
Air Force, U.S., 29, 39
Airlock, 110–111
American Psychiatric Association, 37
American Psychological Association, ix
    Division of Engineering Psychology, 4
    Monitor on Psychology, vii
Ames Research Center, Moffett Field, California, 67, 74
Amundsen, Roald, 54
Analog environments, 13–14, 21, 105
    advantages and limitations, 47–49, 51, 55, 58
    defined, 47–48
    simulations and, 48, 51, 71, 201
    types, 49, 57, 62–76
Analogs. *See* space analogs
Anecdotal evidence, 149
Animals, 17–19
    retrieval of, 19
Annes, C. A., 36
Ansari X Prize, 104
Antarctic Plateau, 73
Antarctica, 20–21, 29, 51, 53–55, 71–73, 126, 127, 196
Apollo, 11–12, 20, 24, 26–27, 29, 38, 65–66, 81–82, 104, 108, 146, 177–181
    Apollo 11, 11–12, 65–66
    Apollo 13, 65–66
    Apollo-Soyuz Test Project, 20, 38, 177–178, 180–182
    Applications Program, 20

Aquarius (underwater habitat), 70
Archimedes, 48
Architecture, 15, 125
Arctic, 65, 71, 74–75
Armstrong, Neil, 48
Asteroids, 126, 130, 141, 175
Astronaut selection, 1, 6, 10, 33–36, 38, 40–41, 145–146, 198
    changing procedures, 34–37, 202
    criteria for, 58–59, 66–67, 76–78
    history of, 17, 33
    interviews, 38
    psychiatry's role in, 36
    psychological testing, 34–38, 202
Astronaut training, 24–25, 38, 181, 185
    cross-cultural, 184–185
    families and, 42
    psychological and interpersonal, 42
"Astronautrix," 25
Astronauts
    attitudes toward psychology, 25, 31
    Canadian, 180
    cultural characteristics of, 177
    European, 179–180
    as exceptional individuals, 192
    female, in international space missions, 125, 129
    gender, 56, 68, 71, 75, 125, 128, 130–132, 140–141
    German, 179
    Japanese, 180
    Korean, 180
    Mercury, 4, 24, 27, 34
    Mir, 26
    multinational, 49, 55, 77
    preparation of, for international missions, 75, 77
    women, 25, 38–39
"Astro-nettes," 25
Australia, 146
Australian National Antarctic Research Expeditions, 21
Aviation human factors, 4, 16

Baker (monkey), 18–19
Bales, R. F., 112, 116
Bales Interaction Analysis, 112, 116
Balloons, high altitude, 17
Bardi, A., 190
Baskin, Pamela, 13, 205
Bates Harris, Ruth. *See* Harris, Ruth Bates
Baum, Andrew, 120
Behavioral health, 7, 14, 22, 44
    conference on new directions in, 31
    isolation, 49, 52, 54–55, 58–59, 61, 67, 70–73, 75, 86
    NASA, perspectives on, 6–7, 10, 198
    personal growth, 86
    psychological, 80, 86–88, 100–101
Behavioral Health and Performance Program, 10, 30, 44, 183
Belgica, 54
Bell, Paul A., 120
Bergerac, Cyrano de, 144
Berry, Charles A., 8
Bigelow, Robert, 104, 119
Bigelow Aerospace, 119
"Big-eye," 71
Bioastronautics Critical Path Roadmap, 17, 29, 43
Bishop, Sheryl, 13, 132, 140, 202, 205–206
Boyd, Jennifer, 13, 206. *See also* Ritsher, Jennifer Boyd
Brady, Joseph V., 7, 18–19
Braun, Wernher von. *See* von Braun, Wernher
Brislin, R. W., 187
Bulgaria, 179
Burrough, Bryan, 26, 181
Bush, George W., 126, 174
Bystander intervention, 121
California Institute of Integral Studies (CIIS), ix
Canada, 74, 146, 157, 174
Canadian Arctic, 74
Canadian Space Agency (CSA), 74

Subject Index

Capsule habitat, 53, 55, 68–70
Cassel, Lindi, 14, 199, 206–207
Cernan, Eugene, 26
Challenger (Royal Navy), 63
Challenger (Space Shuttle), 16, 41, 47, 197
Chamber research, 48, 51–53, 55, 59, 66–68
Chapanis, Alphonse, 5, 7
Chernobyl, 5
Chile, 71
Choosing the Right Stuff (Patricia Santy), 35
Chrétien, Jean-Loup, 149–150, 179
Civilian simulated spaceflight, 15, 31
Claremont McKenna College Aerospace Psychology Laboratory, 109, 122
Clinton, William Jefferson, 20
Cochran, Jackie, 39
Cognition, 10, 14, 26, 44, 168, 182
Cohesion, 131–133, 136, 154
    and additive tasks, 135, 137
    and disjunctive tasks, 135, 137
    interpersonal, 56, 58, 67, 129, 134, 137–139, 200
    task, 14, 129, 134, 137, 200
Cold War, 144, 154, 178
Collectivism-individualism, 159, 185, 188, 190
Collins, Eileen, 41, 130
Collins, Michael, 25
Columbia (Space Shuttle), 13, 16, 47, 84, 177, 179, 197
Committee on Creating a Vision for Space Medicine During Travel Beyond Earth Orbit, 11
Communication, 49, 50, 62, 152
    nonverbal cues, 127–128
    organizational roles, 127
    personality, 127
    team, 55–56, 59–60, 68, 75
Computers, 5, 12, 44, 202
Concordia (Antarctic station), 73–74, 196
Confidentiality, 25
Confinement, 49, 52, 54–55, 58–59, 61, 66–67, 69, 71–73, 75, 125, 169
    laboratories, 66–68

Conflict
    interpersonal, 132, 140
    resolution, 50
Connery, H., 71
Connors, Mary M., 9, 11, 33
Conrad, Pete, 108
Conscientiousness, 60
Control group, 111, 114–115
Cook, Captain James, 63
Cooper, H. S. F., 27
Copernicus, Nicolaus, 48
Coping strategies, 52, 57, 59–60, 70, 73, 160, 166–167, 171, 202
    Accept Responsibility, 161, 167, 171
    Compartmentalization, 161
    Confrontation, 161, 167
    Denial, 161, 167, 171–172
    Distancing, 161
    Endurance/Obedience/Effort, 161
    Escape/Avoidance, 161, 167, 171–172
    Planful Problem Solving, 161, 167, 169
    Positive Reappraisal, 161
    Seeking Social Support, 161, 167, 169
    Self-Control, 161
    Supernatural Protection, 160–161, 167, 171
Cosmonauts
    astronauts and, 38–39
    diaries, use of, 32, 50
    as exceptional individuals, 192
    guest. See *Interkosmos*
    in joint missions with astronauts, 177
    mental health of, 29
    space station, 26
Costa, P. T., Jr., 60
Countermeasures, 10, 198–199, 202
Crew composition, 1, 3, 11–12, 125, 127, 147–148, 195
    culture, 63–64, 68, 128, 177, 201
    gender, 14, 56, 71, 75, 130–132, 137–140, 180, 201
    size, 1, 11–12, 127

Crew dynamics, 126, 201
    cohesion, 56, 58, 67, 131–133,
        136–139
    communication, 49–50, 62
    composition, 77, 125, 130
    conflict, 128, 132, 140
    decision making, 50, 62, 128, 130
    group size, 127–128
    leadership, 51, 56, 59, 137, 139
Crew Earth Observation, 79–101
Crews, international, 75, 77
Critical incidents, 187
    cross-cultural, in space, 68
Cross-cultural issues, 14, 179, 182, 193
Cross-cultural psychology, 177–178, 183
    definition of, 183–184
    origins of, 178
    in space missions, 202
    value of international observations in
        space for, 193–194
Crowding, 106, 118
Cuba, 179
Cultural characteristics, 38, 125, 188–192
    American, 177, 184–185, 188, 191
    Japanese, 188–189, 191–192
    Russian, 188–191
Culture, 3, 22, 152–155, 159, 173, 181–183,
    187–190
    conflict, 11, 27, 128, 197
    definition of, 183
    dimensions of, 182, 187–188, 191–192
    effects on spaceflight, 197, 200
    incidents during international
        spaceflights, 128
    subjective and objective, 185, 187
Culture assimilator, 14, 177, 184–185
    description of, 184
    in preparing astronauts for
        international space missions,
        184–185
    in promoting anticipatory culture
        learning, 185
    specific aspects of, 184–185

Cunningham, Walter, 28
da Vinci, Leonardo, 155
Daedalus and Icarus, 48, 144
Darley, John, 121
Data analysis, 115
    index of amicability, 115
Database, behavioral, 44
de Bergerac, Cyrano. *See* Bergerac, Cyrano de
Decision aids, computer, 13
Delta Clipper, 108
Density intensity hypothesis, 112, 116
Devon Island, 74
Diagnostic and Statistical Manual III, 37
Dialogue between psychologists and
    historians, vii
Discovery (Space Shuttle), 125–126
Discovery Channel Canada, 74
Division of Engineering Psychology,
    American Psychological Association, 4
Division of Polar Programs, National Science
    Foundation, 21
Dome C, 73
Dominion Explorers, 65
Douglas, William K., 28
Draguns, Juris G., 14, 207–208
Dudley-Rowley, M., 65
Dynamism
    Confucian, 188
    Western, 188
Earth observations
    images, 80–85, 87–100
    orbit, 81–82, 84, 89–90, 93–94, 97, 101
    photographs, vii, 80–85, 87–91, 94,
        97–101
Endeavour (English expedition), 63
Enos (chimpanzee), 19
Environmental factors, 19
    human-environment interfaces, 55, 77
    human-human interfaces, 55, 77
    human-technology interfaces, 55
    inaccessibility, 50
    lack of communication, 50
    lack of real-time support, 50

Subject Index

Environments, 149
  analog, 13, 21, 47–77
  laboratory, 49–54, 68–70, 76
  simulators/simulations, 48, 51, 71
  spaceflight, 23, 56
Equal Opportunity Employment Program, 40. *See also* Affirmative Action
Equipment, 149
  70-mm Hasselblad, 82–83
  250-mm lens, 82, 84
  400-mm lens, 84
  800-mm lens, 84, 90–92, 97–99, 101
  Kodak 460 DSC, 84
  Kodak 760 DSC, 85
Ergonomics. *See* human factors
Europe, 146
European Project for Ice Coring in Antarctica (EPICA), 73
European Space Agency (ESA), 15, 68, 74, 196–197
EVA (extravehicular activity), 4, 80, 87, 90, 93, 95, 150
Experimental design
  civilian participants, 111
  clothing, 111
  controlled variables, 111
  crewmembers, 111
  eating, 111
  hygiene, 112
  hypotheses, 112
  measurement techniques, 112
  Mission Control, 112
  noise, 112
  observing participants, 112
  outside viewing, 112
  participant recruitment, 111
  placebo training, 111
  preflight training, 111
  sleeping, 110
  waste management, 110
Experimental group, 111
Experiments, 110–111, 116, 152–153
Exploration Medical Capability Program, 30

Extravehicular activity (EVA), 4, 80, 87, 90, 93, 95, 150
Extreme environments, 137
Extreme human environments
  space analogs, 47–77
  space habitats, 47
Extreme unusual environments (EUEs), 68–69
Facial expressions, 45
Fatigue, 15, 56, 70, 72
Fedderson, William E., 8
"Feminauts," 25
Fiedler, Edna R., 198, 208
Fiedler, F. E., 184
Finney, Ben, 63–64
Flaherty, Bernard, 8
Flashline Mars Arctic Research Station (FMARS), 74–75
Fletcher, James, 40
Flickenger, Don, 38
Flight duration, 110, 116, 162, 171
  long-duration, 116, 150, 153–154, 165, 167, 169, 173
  short-duration, 116, 171
Flynn, Christopher, 27
Focusing events, 45
Foushee, H. Clay, 12
France, 73, 146, 157, 179
Freedman, J. L., 106, 118
French Navy, 64
Freud, Sigmund, 155
Friendship 7, 81
Frozen Sea, 65–66
Gagarin, Yuri A., 48, 103, 119, 178–179, 189
Galarza, Laura, 37
Galilei, Galileo, 48
Gazenko, Oleg, 32
Gemini, 11–12, 20, 36, 82
Gender, 14, 24, 125–126, 128, 130, 137–138, 144, 162, 164–165, 167
  differences, 36, 38–41, 128–130, 200–201
  heterogeneity vs. homogeneity, 14, 121, 138–141
  vs. sex, 128

239

Genesis I and II, 120
GENMOD, 90–91, 95
Genovese, Kitty, 121
Germany, 146, 157, 179
Glenn, John H., 7, 81, 119
GLIMMIX, 90, 95
Goddard, Robert, 48
Golden Gate Bridge, 98
Goldin, Dan, 40
Greece, 185
Greene, Thomas C., 120
Group dynamics, 4, 7, 54, 59, 67, 77, 114
Group for Psychological Support (Soviet Union), 33
Groups
    expeditionary teams, 50, 53, 62, 68
    historical expeditions, 50, 54, 62–66, 77
    military teams, 51
    mountaineering expeditions, 51, 62
    multicultural, 49, 77
    multinational, 49, 55, 77
    polar expeditions, 51, 54–57, 59, 61–62, 65–66, 71
    real world, 49–52, 77
Guest cosmonaut program, 12, 38. *See also Interkosmos*
Gunderson, E. K. Eric, 21, 54
Halvorson, Todd, 28
Ham (chimpanzee), 19
Ham radio, 42
Happiness, 103
Harris, Bernard, 67
Harris, Philip R., 6
Harris, Ruth Bates, 40
Harrison, Albert A., ix, 14, 54, 106, 127–128, 198, 208–209
Harvey, Brian, 26
Haughton Crater, 74
Haughton-Mars Project (HMP), 74
Helmreich, Robert L., 6, 12, 32, 44, 60
Helms, Susan, 28
Hewrmaszewski, Miroslaw, 179
High fidelity, 107

History
    of psychological testing in space program, 4
    of the psychology of space exploration, vii, x
Hofstede, G. J., 187–189
Holland, Al, 26–27, 37
Hong Kong, 178
Human factors, 1, 4–5, 8, 16
    attitudes toward, 23, 43
    conceptual aspects of, 11
Human Health and Countermeasures Program, 30
Human Research Program, 17, 30, 43
    Integrated Research Plan, 43
    Requirements Documents, 43
Icarus and Daedalus. *See* Daedalus and Icarus
Iliad, 54
Impression management, 156
Index of amicability, 115
India, 146, 179
Individualism-collectivism, 159, 185, 188, 190
Institute of Geography, 83
Interdisciplinary collaborations, x
Interkosmos, 146, 149, 157, 164, 172, 177, 179–180. *See also* guest cosmonaut program
International Academy of Astronautics (IAA), ix, 181, 209, 211, 213, 215
International Association for Cross-Cultural Psychology, 178
International Ergonomics Association, 4
International Geophysical Year (IGY), 20
    of 1956–1957, 54, 71
    of 1957–1958, 65–66
International rivalry, 144
    Cold War, 144, 154
    space race, 144, 174–175
International Space Station (ISS), 11–13, 20, 22, 29, 42, 45, 58, 80–101, 125, 130–131, 139, 141, 143, 146–147, 152–154, 157, 174, 177, 180, 196–197
    Medical Project, 30

International Trans-Antarctic, 66
Interpersonal, 148, 154, 160, 169, 172–173
    interaction, 10, 86, 110, 140, 149, 153, 159, 160–162, 164–168, 172, 182
    relationships, 1, 10, 27, 159, 160, 172, 174
Isolation, 8, 15, 26, 49, 52, 54–55, 58–59, 61, 67, 70–73, 75, 86, 125
Italy, 73, 179, 185
Jahn, Sigmund, 179
Japan, 146, 157, 179, 185, 188–189
Job analysis, 38
Johnson Space Center, Houston, Texas, 21, 82, 183
Jones, David R., 36
Kaminski, Megan A., 14, 200, 209
Kanas, Nick, 2, 8, 32, 137
Kass, J., 2
Kass, R., 2
Kennedy, John F., 39
Key Largo, 70
Kon-Tiki, 62
Korolev, Sergey, 48
Krikalev, Sergei, 99
Kring, Jason, 14, 181–183, 200, 209–210
Kubis, Joseph, 8
Kushner, K., 187
Lady Franklin Bay, 65–66
Laika, 18
Latané, Bibb, 121
Launius, Roger, 23
Lebedev, Valentin, 131, 153
Lehman, D. R., 192
Leon, Gloria, 15, 210–211
Lindsley, Donald B., 9
Linenger, Jerry, 29, 81, 151, 154
Linguistic Inquiry and Word Count (LIWC), 160–161, 167–168, 172
Lomov, B. F., 32
Long-duration missions, 47, 61–62, 69, 154, 173, 183, 196
Long-duration spaceflight (LDSF), 126, 129, 137–138

Long-term vs. short-term time orientation, 62, 69, 75
Lovelace, Randall, 39
Lovelace Clinic, 34, 39
    Mercury astronauts, 39, 129
    women tested at, 39, 129
Low, George, 24
Lowell, Percival, 48
Lucid, Shannon, 150–151, 169
Lugg, Desmond, 21
Lunar missions, 11–12, 20, 30, 42–43, 86, 104–105, 119, 126, 147, 174–175, 196, 198–199. *See also* Apollo
Management, 7, 55, 177, 181–182
Mars, 126, 141
    Flashline Mars Arctic Research Station (FMARS), 74–75
    Haughton-Mars Project (HMP), 74
    human missions, 11–13, 16, 30, 42–43, 58, 73, 80, 86, 101, 104–105, 119, 123, 126, 130, 141, 174–175, 196, 199–202
    Mars Desert Research Station (MDRS), 75
Mars Society, The, 74–76
Marshall Space Flight Center, Huntsville, Alabama, 67
Marxism, 155
Masculinity-femininity, 188, 190
    ranks in, for Japan, Russia, and the United States, 188–189
McCrae, R. R., 60
McDonnell Douglas Aerospace Corporation, 107–108, 113
McKinley, Mt., 99
McLaughlin, Edward J., 8
Measurement, 10
    "big-eye," 71
    critical incidents, 187
    depression, 71–72
    group fission, 58
    group fusion, 58
    hostility, 71

impaired cognition, 71
personal diaries, 50, 54
psychological, 153, 157–160, 164
psychophysiological, 52
psychosocial, 49–50, 53, 55, 68, 71–72, 76
salutogenic, 72
self-efficacy, 73
self-reliance, 73
winter-over syndrome, 71
Medical Sciences Division, NASA, 37
Meister, David, 11
Merbold, Ulf, 179
Mercury, Project, 4, 11–12, 17, 19–20, 24, 27, 34–35, 38–39, 82, 104, 129, 156
Merritt, A., 189
Meyers, David G., 122
Microgravity, 15, 19, 47–48
Milgram, Stanley, 121
Mir Space Station, 1, 11–12, 17, 20, 22, 26, 29, 31–33, 45, 83, 130, 143, 150–151, 153–154, 169, 171, 173, 177, 181, 198
Mission Control, 3, 26–27, 51, 68, 70, 77, 126, 128, 150, 152, 173
Mission phase, 158, 164–169
in-flight, 164–166, 168–169
postflight, 164–166, 169
preflight, 164–166, 168
Missions
Apollo, 11–12, 20, 24, 26–27, 29, 38, 65–66, 81–82, 104, 108, 146, 177–181
asteroid, 126, 130, 141, 175
extended duration, 1, 9, 11–13, 43
Gemini, 11–12, 20, 36, 82
Genesis, 120
Interkosmos, 146, 149, 157, 164, 172, 177, 179–180
International Space Station (ISS), 11–13, 20, 22, 29, 42, 45, 58, 80–101, 125, 130–131, 139, 141, 143, 146–147, 152–154, 157, 174, 177, 180, 196–197

lunar, 11–12, 20, 30, 42–43, 86, 104–105, 119, 126, 147, 174–175, 196, 198–199
Mars, 11–13, 16, 30, 42–43, 58, 73, 80, 86, 101, 104–105, 119, 123, 126, 130, 141, 174–175, 196, 199–202
Mercury, 4, 11–12, 17, 19–20, 24, 27, 34–35, 38–39, 82, 104, 129, 156
Mir, 1, 11–12, 17, 20, 22, 26, 29, 31–33, 45, 83, 130, 143, 150–151, 153–154, 169, 171, 173, 177, 181, 198
Salyut, 7, 26, 32, 65–66, 130–131, 149–150, 177
Skylab, 12, 20, 27–28, 32, 82
Space Shuttle, 9, 11–12, 20, 32, 42, 48, 81–82, 87, 125, 143, 150–151, 154, 173, 180–181
Spacelab, 177, 179–180
Vostok, 129
Mistacoba experiment, 74
Mitchell, Edgar, 29
Modules, 146
Monitor on Psychology, vii
Moon missions. *See* Lunar missions
Mount, Frances, 15
Mt. McKinley. *See* McKinley, Mt.
Mullin, C. S., 54, 71
Multinationality, 143–175
American-Russian team, 143, 180
differences in national origin, 148–149, 154, 157–158, 162
foreign and compatriot visitors, 146–147, 149–151, 153, 172
host nationality (American, Russian), 143, 172–173
host-guest relationship, 143–144, 146–147, 150, 152–153, 162, 166, 171–173
intercultural issues, 128, 143, 148, 153
mixed-nationality crews, 128, 143, 146–150, 157, 172
Musson, David, 44

Subject Index

NASA (National Aeronautics and Space Administration), 2, 6, 11, 17, 21, 35, 37, 42, 129, 150, 156–157, 162, 172, 174
    Ames Research Center, 67, 74
    behavioral health and, 6–7, 10, 198
    Communications Support Services Center, ix
    employment of women by, 38–41, 125–126, 129–130
    equal opportunity employment in, 39–41
    Extreme Environment Mission Operation (NEEMO), 70
    History Division, ix, 178
    Johnson Space Center, 21, 82
    Marshall Space Flight Center, 67
    Medical Sciences Division, 37
    psychology, role in, 6–7, 16–17, 20, 22–25, 32–33, 35, 39, 44–45, 150, 198
    public relations and, 2, 24, 35
    Weightless Environmental Training Facility (WET-F), 48
NASA-*Mir*, 81, 83
National Academy of Sciences, 9, 11, 22, 40, 74
National Commission on Space, 55
National Geographic Society, 74
National Oceanic and Atmospheric Administration (NOAA), 70
National Research Council, 74
    Committee on Space Biology and Medicine, 9–11
National Science Board, 55
National Science Foundation, 6, 21
National Space Biomedical Research Institute, 17, 30–31, 37, 43
National Undersea Research Center (NURC), 70
Nautilus (submarine), 69
Negative interpersonal interactions, 109, 121, 123, 152, 156, 160, 164–166, 168–169, 172–173

NEO-Personality Inventory, 60
Neuroticism, 36, 60–61
New way of thinking, 118
    about equipment, 119
    about spacefarers, 121
Newton, Isaac, 48
Nicollier, Claude, 179
Nonprofessional spacefarers, 151, 156
North Pole, 62
Nutrition, 5
Obama, Barack H., 126, 174
Obedience, 121, 161
Oberth, Hermann, 48
Odum, Floyd, 39
Odyssey, 54, 179
Olson, Valerie A., 13, 211
One-way window, 110–111
Openness to experience, 60
Overview effects, 29
Palinkas, Lawrence, 7, 53, 58, 72
Palmer Station, 58
Payload specialists, 151, 157
Performance, 2, 9, 26, 30, 147, 170, 173, 198
    autonomous decision-making, 50
    command and control, 51
    conflict resolution, 50
    group dynamics, 3, 10, 54, 59, 67, 77
    leadership, 51, 56, 59
    problem-solving, 50
    selection, 58–59, 66–67, 76–77
    self-monitoring, 50, 77
    self-regulation, 50
    support, 47, 49–50, 58–59, 69, 73, 77
    team communication, 49–50, 62
    team cooperation, 55–56, 68
    team coordination, 55–56
    training, 47, 68, 70, 77
Perrow, Charles, 23
Personality differences, 53, 56–57, 60–61, 127, 131, 151, 154–155, 174
Petrov, B. N., 32
Pettit, Don, 97
Placebo, 111, 115–116

Positive psychology, 44, 147
Positive reactions, 80, 86, 101, 147, 151, 153, 160–161, 164–165, 167–170, 172, 199
    changes in values, 199
    personal growth, 86
Power distance, 188, 190
Privacy, 22, 58, 125, 181
Progress, 87, 174
Psychiatrists, vii, 1, 23, 37, 42
Psychological research, vii
    confidentiality, need for, 25
    involving astronauts in, 31
    unobtrusive, 45
Psychological support, 1, 13, 17, 41–42
Psychological testing, 202
    applicant screening, 198
    history of, in space program, 4
Psychologists, vii, 1, 42, 44, 147
Psychology, 15, 42
    astronauts as research subjects, 149
    personality, social, and organizational, 1, 6
    positive, 44, 199–200
Psychophysiological Aspects of Space Flight (Bernard Flaherty), 8
Psychophysiology, 144, 147
Psychosocial adjustment, 1, 17
Psychosocial factors, 125
    confinement, 49, 52, 54–55, 58–59, 61, 66–67, 69, 71–73, 75, 126
    isolation, 49, 52, 54–55, 58–59, 61, 67, 70–73, 75, 126
Psychotherapy, 202
Public relations, 24–25, 156
Reagan, Ronald W., 20
"Red hands" disease, 149, 179
Remek, Vladimir, 149–151, 179
Research, 7, 10, 13, 16, 25, 36, 43, 45, 152
    archived, 81
    collection, 100
    data, 80–82, 86–87, 89–91, 97, 100–101
Ride, Sally, 41, 130

Right Stuff, The (Tom Wolfe)/right stuff, 17, 23–33, 41, 60, 61, 119, 198
Ritsher, Jennifer Boyd, 152, 180, 189, 202. *See also* Boyd, Jennifer
Robinson, Julie A., 13, 200, 211–212
Robotics, 13
Rogers, Buck, 8
Romania, 179
Royal Navy, 63
Russia, 20, 146–147, 149–154, 156–158, 162, 164–165, 167, 169–174
Russian Academy of Sciences, 83
Russian Institute for Space Research (IKI), 74
Russian space program, 6, 26–27, 143, 145–152, 154, 157, 159, 162, 164–165, 168, 170–174
    NASA and, views of psychology, 32–33
    space race and, 25, 32, 144–145, 174–175
    space stations and, 154
Russian space stations
    Mir, 1, 11–12, 17, 20, 22, 26, 29, 31–33, 45, 83, 130, 143, 150–151, 153–154, 169, 171, 173, 177, 181, 198
    Salyut, 7, 26, 32, 65–66, 130–131, 149–150, 177
Rutan, Burt L., 104
Ryan, Craig, 7
Safe Passage: Astronaut Care for Exploration Missions (John R. Ball and Charles H. Evans, Jr., eds.), 11, 17, 22, 43, 49
Salutogenic experiences, 72, 80, 86, 101, 147, 199–200
Salyut, 32, 130, 149–150, 177
    Salyut 7, 26, 65–66, 131
Samaltedinov, I., 2
Samsonov, N. D., 32
Sandal, Gro M., 15, 59, 71, 116, 212
Santy, Patricia R., 6–7, 35–37, 128
São Paulo, Brazil, 97
Savitskaya, Svetlana, 130–131
Scaled Composites, LLC, 103–104

Schmitt, Harrison, 28
Scholarship, criteria for, vii
Schwartz, S. H., 158–159, 190
Scientific Committee for Antarctic Research, 21
Scott, Robert F., 54
Scripps Institute of Oceanography, 64
Selection, 1, 6, 13, 17, 34–37, 40–41, 43, 58–59, 66–67, 76–77, 145–146, 201
Self-report studies, 148, 153–154, 156
Sells, Saul B., 8
Sensitivity, cultural, 14, 182, 192
Sensory deprivation, 66–67, 161
SETI Institute, ix
Sex, 148, 158
    vs. gender, 128
Shepanek, Mark, 25
Shepard, Alan, 81, 178
Shuttle. *See* Space Shuttle
Shuttle-*Mir* program, 29, 83, 150–151, 154, 173
Simulated spaceflight experiment, 6, 13–14, 17, 20, 123, 196
    Delta Clipper, 108
    McDonnell Douglas Aerospace, 108
    research purpose, 108
    spaceflight simulator, 108, 148, 153
Simulation of Flight of International Crew on Space Station (SFINCSS), 132
Single-stage-to-orbit rocket (SSTO), 108
Sipes, Walter, 42
Skylab, 12, 20, 27–28, 32, 82
Slack, Kelley J., 13, 212–213
Sleep, 26, 30–31, 116, 149
Snow, C. P., vii
Social interaction, 167–168
    research with N of 1, 193
Social perception, 145, 152
Social relations, 155, 159–161, 164–165, 167–174
    Affiliative Trust/Mistrust, 160, 164, 169
    Negative Intimacy, 160, 164–166, 169
    Positive Intimacy, 160, 164–166

Society for Industrial and Organizational Psychologists, 38
South Africa, 179
South Pole, 29
Soviet *Interkosmos* program. *See Interkosmos*
Soviet space program. *See* Russian space program
Soviet Union, 145, 146–147, 149, 156–157. *See also* Russia
Soyuz, 21, 26, 42, 47, 87, 146, 149, 180
    Apollo-Soyuz Test Project, 20, 38, 177–178, 180–182
Space Age, vii, 44, 143–144
Space analogs, 13, 20, 47–77, 106–107, 148, 153
    Antarctic stations, 20–21, 71–73, 126–127, 132, 139, 140, 153
    armored vehicles, 66
    artificial environments, 51
    caving expeditions, 55, 62
    chamber studies, 48, 51–53, 55, 59, 66–68
    civilian and military aviation, 56
    constructed environments, 51, 67
    environmental chambers, 48, 51–53, 55, 59, 66–68
    extreme environments, 47, 49, 52–54, 56–57, 72, 76, 132, 137
    extreme unusual environments (EUEs), 69
    hybrids, 53, 74
    hyperbaric chambers, 59, 68, 196
    in situ studies, 49, 52, 196
    laboratory studies, 49, 51–52, 54, 76
    long-duration, 47, 61–62, 69
    natural environments, 50, 52
    naval ships, 63, 140
    oceanographic research vessels, 64–65
    radiation profiles, 48
    reduced-gravity environments, 48
    short-duration, 62, 69, 75
    simulations, 6, 8, 14, 17, 20, 24, 106–107, 148, 153, 196

submarines, 20, 55, 69, 127
submersible habitats, 69–70, 153
terrestrial substitutes, 47
"tuna can" habitats, 69
underwater capsules, 55, 153
winter-overers, 61, 72, 153
Space Biomedical Research Institute, 37
"Space girls," 25
Space hotels, 104, 119
Space Human Factors and Habitability Program, 30
Space medicine, 11
Space Radiation Program, 30
Space Science Board, 55
Space Shuttle, 9, 11–12, 20, 32, 42, 48, 81–82, 87, 125, 143, 150–151, 154, 173, 180–181
    Challenger, 16, 41, 47, 197
    Columbia, 13, 16, 47, 84, 177, 179, 197
    Discovery, 125–126
Space Shuttle Earth Observation (SSEOP), 82–84
Space stations, 1, 9, 12, 20, 23, 30, 32, 41, 145, 149, 151
    astronauts and, 156
    cosmonauts and, 156
    International Space Station (ISS), 11–13, 20, 22, 29, 42, 45, 58, 80–101, 125, 130–131, 139, 141, 143, 146–147, 152–154, 157, 174, 177, 180, 196–197
    Mir, 83, 143, 150–151, 153–154, 169, 171, 173, 177
    Salyut, 7, 26, 32, 65–66, 130–131, 149–150, 177
    Skylab, 12, 20, 27–28, 32, 82
Space tourism, 1, 14, 103–105, 108, 116–117, 121–122, 180, 197
Space tourists, 12, 14, 38, 105, 119, 122, 156
Space.com, 28
Spaceflight simulators, 108, 122
Spacelab, 177, 179–180
Speech, stress and, 45

Stanford prison experiment, 121–122
Stereotypes, 180
Stress, 2–3, 9, 34, 43, 52–53, 56–57, 60, 70, 73, 130, 147, 148, 169, 171, 173, 189
    workload, 127
Stressors, 10, 43, 56–57, 70, 147–148, 173
    career, 2
    environmental, 169
    psychological, 147, 171
STS-73, 84, 186
Submarines, 55, 69, 127
Submersible habitats, 20, 69–70
Suedfeld, Peter, 7, 14, 57, 60, 199, 213–214
Sunnyvale Conference, 21
Superordinate goals, 180
Syria, 146, 179
Systems, 9, 11, 13, 31, 152
Takata, T., 192
Teams, 3, 147, 150–151, 154
    mixed-gender, 71, 128–129, 137–138
Teamwork, 57
Technology, 13, 144–145, 154
Tereshkova, Valentina, 125, 129, 130
Terra Nova Expedition, 65–66
Test pilots, 17, 33–34, 146, 148
Thagard, Norman, 150–151
Thematic Content Analysis (TCA), 144, 148, 155–167
    autobiographies, 144
    databases, 144, 156
    media interviews, 144, 148, 156, 160
    memoirs, 144, 147–148, 156, 168–169
    oral history interviews, 144, 150–151, 154, 156, 160, 169
    quantitative data, 148, 155–156
Three Mile Island, 5
Tito, Dennis A., 103
Tourism, space, 1, 14, 103–105, 108, 116–117, 121–122, 180, 197
Tourists, space, 12, 14, 38, 105, 119, 122, 156
Training, 6, 10, 17, 47, 68, 70, 77, 147, 149, 151, 153–154, 173, 181–182, 184, 189
Trenchard, Mike, 13, 214

Triandis, Harry C., 184–186
Tsiolkovsky, Konstantin, 48
Two cultures, vii
Type A personality, 60
UCLA (University of California, Los Angeles), 9
Uncertainty avoidance, 152, 188
    as a cultural dimension, 188
    definition of, 188–189
United Kingdom, 179
United States Exploring Expedition, 64
University of Hawaii, 63
University of North Carolina at Wilmington, 70
U.S. Marine Corps, 74
U.S. National Academy of Sciences, 9, 11, 22, 40, 74
U.S. National Oceanographic and Atmospheric Administration (NOAA), 70
U.S. Naval Support Force Antarctica, 139
Utah, 75
Vakoch, Douglas A., 214–215
Value hierarchies, 158–159, 162, 164, 170
    Achievement, 159, 162–164, 170–171
    Benevolence, 159
    Conformity, 159, 162–164, 168, 171, 172
    Hedonism (enjoyment), 159, 162, 164, 168, 172
    Power, 152, 159, 164, 169
    Security, 152, 159, 162, 171–172
    Self-Direction, 159, 162, 164, 169
    Spirituality, 159–160, 162, 164, 168, 171
    Stimulation, 159, 164, 168
    Tradition, 159
    Universalism, 159, 162, 172
Value orientation, 152, 174, 187, 190–191
    cultural dimensions of, 152, 164, 170, 187
    research on, 177
Values, 144, 147, 162, 168–170, 173–174, 200
    differences between Russians and Americans, 152, 159, 162, 164
    in multinational space crews, 162, 164–175, 197
    national differences in, 197
Vander Ark, Steven, 42
Verne, Jules, 48, 144
Vietnam, 179
Vinci, Leonardo da. *See* da Vinci, Leonardo
Virgin Galactic, 104
Virgin Islands, 70
Vision for Space Exploration, 85, 126
Voas, Robert, 21, 24
von Braun, Wernher, 18, 48
Vostok 6, 129
Wayne, John, 25
Weick, Karl, 27
Weightless Environmental Training Facility (WET-F), 48
Weightlessness, 5, 15, 19
Weitekamp, Margaret A., 7–8, 38, 39
Wells, H. G., 48
White, Frank, 29
Whitson, Peggy, 12–13, 130
Wichman, Harvey A., 14, 197, 215–216
Wilk, Kasia E., 14, 199, 216
Williams, Jeff, 98
Williams, Sunni, 12
Willis, Kimberly J., 13, 216–217
Windows, 14, 22, 28, 80, 85, 89, 92, 99, 113
    of opportunity, 31
Winter-over syndrome, 71
WISE (Women in Space Earlier), 38
Wolfe, Tom, 41
Women
    as astronauts, 25, 38–39, 125, 129, 145–146
    Lovelace's Women in Space Program, 129
    NASA, employment by, 38–41, 125, 129–130
    qualifications of, 129
    in space, 125, 129

Women in Combat Task Force Study Group, 138
Woods, Joanna, 21
Woolford, Barbara, 15
Work, 2, 4–5, 28, 149–154, 161, 168–169, 171
Wouters, F., 71
Wrangel Island, 65–66
Wright Air Development Center, 34
Wright brothers, 4, 48
Wylie Laboratories, 42
Zedekar, Raymond, 34
Zimbardo, Philip, 128
Zubek, J. P., 67

Authors Cited

Afanas'ev, Y., 190
Akins, F. R., 9, 11, 28, 33, 88, 180
Allik, J., 60
Altman, I., 49, 54, 67
Amsbury, D. L., 82
Andersen, D. T., 21
Anderson, C., 55
Anderson, J. R. L., 146
Annes, C. A., 36
Annes, R., 75
Antipov, V., 29
Apt, J., 81
Atkinson, J. W., 156, 160
Ayton, J., 21
Bachelard, C., 88, 132
Baddeley, A. D., 52
Bales, R. F., 112, 116
Ball, J. R., 11, 22, 85
Barber, H., 136
Bardi, A., 190
Barrick, M. R., 60
Barth, T., 13
Baugh, G. S., 138
Baum, A., 120
Baumann, D., 100
Bechtel, R. B., 88
Bell, P. A., 120
Bergan, P. T., 56, 59
Bergen, S., 150
Bernard, H. R., 62–64
Berning, A., 88
Berry, C. A., 8
Berry, J. W., 184
Betzendoerfer, D., 57
Bioastronautics Critical Path Roadmap, 22, 29
Birkland, T. A., 31
Bishop, S. L., 55–56, 62, 65, 75, 130, 132, 140, 182
Boeing Company, 56

Booth, R. J., 161
Bosrom, A., 154
Brady, J. V., 7, 18–19
Brazil, D. M., 139
Brislin, R. W., 184, 187
Brown, W. R., 56
Bunzelek, K., 75
Burgess, C., 18
Burr, R., 55
Burris, J. W., 138
Burrough, B., 26, 145, 181
Butler, C., 150
Caldwell, B., 65
Carless, S. A., 134
Carpenter, F., 41
Cartwright, D., 133
Casutt, M., 178
Cazes, G., 132
Cernan, E., 26
Chamisso, A. von, 64
Chapanis, A., 5–7
Cizman, Z., 45
Clearwater, Y. A., 6, 21, 54, 57, 63, 69, 71, 72, 88, 117
Cleary, P. J., 52
Colquitt, J. A., 130
Committee on Space Biology and Medicine, 10
Connery, H., 71
Connors, M. M., 9, 11, 28, 33, 88, 180
Cooper, H. S. F., 20, 27
Costa, P. T., Jr., 60
Crespo-Richey, J., 100
Cunningham, W., 28
Currie, N. J., 13
Cushner, K., 187
Darley, J. M., 121
Dasen, P. R., 184
Davies, C., 52, 68
Davis, D., 26

249

Davis, S. F., 128
Davison, M., 151, 169
Dawson, J. L. M., 178
Dawson, S., 75
De Paola, C., 134
DeLongis, A., 160
DeRoshia, C. W., 31, 63
Dessinov, L. D., 81, 83
Devine, D. J., 135
Dickerson, P. W., 81, 83
Dinges, D. F., 45
Dion, K. L., 134, 136–137
Dismukes, K., 120
Doi, L. T., 191
Dorrian, J., 45
Douglas, W. K., 28
Doyle, V. F., 197
Dubbs, C., 18
Dudley-Rowley, M., 65
Dukes, W. F., 192
Dunkel-Schetter, C., 160
Dunmore, J., 63–64
Eagly, A. H., 130
Efimov, V. A., 52, 68
Eggins, R., 75
Ellis, A., 130
Emmart, C., 28
Endler, N. S., 131, 140
Endresen, I. M., 56
Engen, M. L. van, 130
Ethier, B., 81
European Space Agency, 74
Evans, C. A., 81–84, 90, 100
Evans, C. H., Jr., 11, 22, 86
Evans, C. R., 136
Faulk, D., 56, 62
Fedderson, W. E., 8
Ferguson, L., 60
Festinger, L., 133
Feustl-Beuchl, J., 180
Fiedler, E. R., 41, 182
Fiedler, F. E., 184
Finney, B., 63–65

Fisher, J. D., 120
Flaherty, B. E., 8, 67
Flynn, C. F., 25, 27, 31, 41
Fogg, L. F., 60
Folkman, S., 160
Foushee, H. C., 12
Francis, M. E., 161
Freedman, J. L., 106, 118
Freud, S., 155
Furniss, T., 147
Galarza, L., 37–38, 182
Gilluly, R. H., 56
Goetzmann, W. H., 63
Goldenstein, S. K., 45
Goodykunst, W., 191
Graen, G. B., 138
Graham, N., 56
Greely, A., 50
Greene, T. C., 120
Grether, W. F., 4, 7
Grigoriev, A. I., 29
Grobler, L. C., 56, 62
Gruen, R., 160
Grund, E. M., 51, 68, 150–151, 154
Grunzke, M. E., 19
Gully, S. M., 135
Gunderson, E. K. E., 20–21, 54–55, 61, 71–72
Gushin, V. I., 2, 32, 52, 68, 150–151, 154, 182
Haakonstad, E., 120
Hall, R. D., 149
Halvorson, T., 28
Harris, B., 67
Harris, P. R., 6
Harrison, A. A., 6, 9, 14, 21–22, 28, 29, 31, 33, 43, 54, 57, 63, 69, 71, 72, 88, 106, 117, 127–128, 128, 180
Harvey, B., 26
Hatfield, B. D., 136
Haythorn, W., 49, 54, 67
Heine, S. J., 192
Helfert, M., 81
Helmreich, R. L., 6, 12, 32, 44, 56–57, 60, 70
Heyerdahl, T., 62

Hochstadt, J., 45
Hofstede, G., 187–189
Hofstede, G. J., 188–189
Holland, A. W., 26, 37–38, 55, 61, 128, 150, 181–182
Hollenbeck, J. R., 130
Holquist, M., 144
Holsti, O. R., 155
Honingfeld, R., 67
House Committee on the Judiciary, 40
Howard, N., 70
Huntoon, C. L., 29, 182
Hysong, S. J., 182
Ihle, E. C., 86, 88, 147
Ilgen, D. R., 130
Ilies, R., 61
Inoue, N., 182
Jahoda, G., 178
James, D., 60
Johannes, B., 182
Johannesen-Schmidt, M. C., 130
Jones, D. R., 36
Judge, T. A., 61
Jurion, S., 132
Kaiser, M. K., 13
Kaltenbach, J. L., 82
Kanas, N., 2, 8, 31–32, 41, 51–52, 56, 68, 86, 88, 126, 128, 131, 137, 140, 147–148, 150–151, 153–154, 182
Kane, R. L., 31
Kanki, B. O., 56
Kass, J., 2–3
Kass, R., 2–3
Kealry, D. J., 182
Kelly, A. D., 31, 148
Kelly, D. R., 133, 147
Killworth, P., 62–64
Kirkcaldy, B. D., 131
Klein, W., 71
Kluger, J., 50
Kocher, T., 57
Kolintchenko, V. A., 52, 68
Kosslyn, S. M., 45

Kozerenko, O. P., 2, 32, 150–151, 154
Kraft, N., 182
Krell, R., 160
Kring, J. P., 181–183
Kubis, J. F., 8
Kumar, P., 148, 152–154
Landers, D. M., 136
Langfred, C. W., 133
Larson, M., 45
Latané, B., 121
Launius, R. D., 24
Lazarus, R. S., 160
Le Scanff, C., 61
Lebedev, V., 50, 131, 149, 153, 173
Lehman, D. R., 192
Lenoir, W. B., 82
Leon, G. R., 71, 182
LePine, J. A., 60, 130
Liddle, D. A., 82
Lieberman, P., 45
Lindsley, D. B., 9
Linenger, J. M., 29, 150–151, 154
Link, M. M., 34
Linklater, E., 63
Lomov, B. F., 32
Lonner, W. J., 178
Looper, L., 128, 181
Lovell, J., 50
Lowe, C. A., 135
Lowman, P. D., Jr., 82
Lucid, S., 150–151, 169
Lugg, D. J., 21
Lulla, K. P., 81, 83
Mallis, M. M., 31, 63
Manzey, D., 41, 68, 86, 126, 128, 131, 140, 147, 153, 182, 197
Marcondes-North, R., 128, 181
Marmar, C. R., 2, 32, 88, 150–151, 154
Mars Institute, The, 74
Mars Society, The, 76
Marsch, S., 57
Mather, S., 45
Matsuzaki, I., 182

McAdams, D. P., 160
McCandless, J. W., 13
McCann, R. S., 13
McCarthy, G. W., 52
McClelland, D. C., 156, 160
McConnell, M. M., 16
McCoy, M. C., 135
McCrae, R. R., 60
McFadden, T. J., 60
McGlinchey, E. L., 45
McKay, C. P., 21, 54, 71, 88, 160
McKay, J. R., 6
McLaughlin, E. J., 8
McQuaid, K., 25, 40
Mears, J. D., 52
Meister, D., 5, 11
Merritt, A., 189
Metaxas, D. N., 45
Meyers, D. G., 122
Miesing, P., 135
Milgram, S., 121
Mitchell, E., 29
Mitchell, T., 184
Morey, A., 45
Moser, R., 56
Mount, F., 15
Mount, M. K., 60
Mullane, M., 145, 151
Mullin, C. S., 54, 72
Musson, D. M., 44, 182
Myasnikov, V. I., 29
National Aeronautics and Space Administration (NASA), 2, 29, 35, 44, 82, 85–86, 89, 129, 145, 150, 156–157, 162, 172, 174, 178
National Commission on Space, 55
National Science Board, 55
Neubert, M. J., 60, 138
Nishida, T., 191
Nolan, P. D., 65
North, R., 155, 182
Oberg, A. R., 32
Oberg, J. E., 32, 179

O'Hehir, F., 60
Oliver, D., 72
Pacros, A., 75
Palinkas, L. A., 7, 20, 29, 51, 53, 55–56, 59, 61, 71–72, 182
Palladino, J. J., 128
Panel on Human Behavior, 151
Panzer, F.-J., 139
Paola, C. De, 134
Patricio, R., 75
Pearce, R., 50
Pennebaker, J. W., 161
Perrow, C. E., 23
Petrov, B. N., 32
Phillips, T., 21
Poortinga, Y. H., 184
Preble, J., 135
Primeau, L., 56
Purvis, B., 52
Rasmussen, J. E., 20, 54, 71
Rawat, N., 75
Raymond, M. W., 56
Reynolds, H. H., 19
Reynolds, K., 75
Rhatigan, J., 100
Rholes, F. H., 19
Ricketson, D. S., 56
Rider, R. L., 45
Ritsher, J. B., 2, 32, 86, 147, 152, 154, 180–183, 189, 202
Robinson, J., 81–84, 90, 100
Rogers, N. L., 45
Rollins, P., 150–151, 169
Rose, R. M., 60
Rosenberger, N. R., 191
Rosnet, E., 61, 132, 139
Rummel, J. D., 21
Runciman, W. B., 57
Ryan, C., 18
Sagal, M., 61
Saimons, V. J., 139
Salnitskiy, V. P., 2, 32, 51, 68, 150–151, 154, 182

Salvendi, G., 15
Samaltedinov, I., 2–3
Samsonov, N. D., 32
Sandal, G. M., 51, 56, 59, 68, 71, 88, 116, 132, 182, 197
Sanders, A., 61
Santy, P. A., 6–7, 24, 32–33, 35–37, 56, 62, 128, 181
Saylor, S. A., 2, 32, 154
Scanff, C. Le, 61
Scheidegger, D., 57
Schjøll, O., 56, 62
Schmidt, H., 28
Schmidt, L., 21
Schwartz, S. H., 158–159, 190
Segall, M. H., 184
Sellen, A., 57
Sells, S. B., 8
Sexner, J. L., 56
Sexton, B., 57
Shappel, S. A., 56
Shayler, D. J., 147, 149
Shayler, M. D., 147
Shepanek, M., 21, 25, 27
Shephard, J. M., 45
Shlapentokh, V., 190–191
Short, P., 31
Shurley, J. T., 72
Shwalb, B. J., 191
Shwalb, D. W., 191
Siebold, G. L., 133
Sipes, W. E., 2, 31, 33, 41, 182–183
Skelly, J., 52
Slayton, D. K., 178
Sled, A., 150–151, 154
Smirnova, T. M., 52
Smith, C. P., 156
Society of Industrial and Organizational Psychology, 38
Space Science Board, 55
Stanton, W., 64
Steel, G. D., 53, 60, 69, 147, 160
Stefanowska, P., 197

Stefansson, V., 50
Steiner, I. D., 134
Stewart, G. L., 60
Stogdill, R. M., 134
Stoker, C., 28
Strange, R., 71
Stuster, J., 50, 88, 127, 132, 139, 140, 147, 182
Suedfeld, P., 7, 27, 29, 44, 51–53, 57, 60, 69, 72, 73, 86, 88, 147, 160, 168–170, 182
Sullivan, K. D., 75, 81
Summit, J. E., 28
Sundaresan, A., 75
Taggar, S., 7
Takata, T., 192
Tarazona, F., 150
Tate-Brown, J., 100
Taylor, A. J. W., 72
Thagard, N. E., 150–151
Thumm, T., 100
Tomi, L., 197
Triandis, H. C., 184–186
Tziner, A., 136
Ursin, H., 51, 55–56, 59, 68, 116
Vaernes, R., 51, 56, 59, 68, 116
van der Walt, J. H., 57
van Engen, M. L., 130
Vander Ark, S. T., 2, 33
Vardi, Y., 136
Vaughan, D., 16
Vecchio, R. P., 139
Venkartamarian, S., 45
Vinokhodova, A. G., 52
Vis, B., 149
Voas, R., 34
Volger, C., 45
von Chamisso, A., 64
Walt, J. H. van der, 57
Warncke, M., 56, 59
Webb, R. K., 57
Weick, K. E., 27
Weiner, B. L., 12, 56
Weiss, D. S., 2, 32, 88, 150–151, 154

Weiszbeck, T., 86, 170
Weitekamp, M. A., 7–8, 18, 25, 38–39, 41
Weybrew, B. B., 69–70
Wharton, R. A., 21
White, F., 29
Whitney, D. J., 135
Whitney, S., 65
Wichman, H. A., 108, 120
Wiebe, R., 160
Wiegmann, D. A., 56
Wilkinson, J., 81
Wilkinson, M. O., 136
Williams, D., 29
Williamson, J. A., 57
Wilmarth, V. R., 82
Wilson, G., 52
Wilson, O., 72
Wolf, D., 151
Wolfe, T., 18, 41
Wood, J., 21, 27
Woolford, B., 13, 15
Wouters, F., 71
Wright, R., 150–151, 169
Yacovone, D. W., 57
Zaccaro, S. J., 134, 135
Zamaletdinov, I. S., 29
Zander, A., 133
Zaninovich, M. G., 155
Zaprisa, T. B., 52
Zedekar, R., 34
Zimmerman, R., 149–150, 173
Zinnes, D. A., 155
Zubek, J. P., 49, 67